U.S.NRC

United States Nuclear Regulatory Commission

Protecting People and the Environment

NUREG/CR-7142
PNNL-21547

Ultrasonic Phased Array Assessment of the Interference Fit and Leak Path of the North Anna Unit 2 Control Rod Drive Mechanism Nozzle 63 with Destructive Validation

Office of Nuclear Regulatory Research

AVAILABILITY OF REFERENCE MATERIALS
IN NRC PUBLICATIONS

NRC Reference Material

As of November 1999, you may electronically access NUREG-series publications and other NRC records at NRC's Public Electronic Reading Room at http://www.nrc.gov/reading-rm.html. Publicly released records include, to name a few, NUREG-series publications; *Federal Register* notices; applicant, licensee, and vendor documents and correspondence; NRC correspondence and internal memoranda; bulletins and information notices; inspection and investigative reports; licensee event reports; and Commission papers and their attachments.

NRC publications in the NUREG series, NRC regulations, and Title 10, "Energy," in the *Code of Federal Regulations* may also be purchased from one of these two sources.
1. The Superintendent of Documents
   U.S. Government Printing Office Mail Stop SSOP
   Washington, DC 20402–0001
   Internet: bookstore.gpo.gov
   Telephone: 202-512-1800
   Fax: 202-512-2250
2. The National Technical Information Service
   Springfield, VA 22161–0002
   www.ntis.gov
   1–800–553–6847 or, locally, 703–605–6000

A single copy of each NRC draft report for comment is available free, to the extent of supply, upon written request as follows:
Address: U.S. Nuclear Regulatory Commission
          Office of Administration
          Publications Branch
          Washington, DC 20555-0001
E-mail: DISTRIBUTION.RESOURCE@NRC.GOV
Facsimile: 301–415–2289

Some publications in the NUREG series that are posted at NRC's Web site address http://www.nrc.gov/reading-rm/doc-collections/nuregs are updated periodically and may differ from the last printed version. Although references to material found on a Web site bear the date the material was accessed, the material available on the date cited may subsequently be removed from the site.

Non-NRC Reference Material

Documents available from public and special technical libraries include all open literature items, such as books, journal articles, transactions, *Federal Register* notices, Federal and State legislation, and congressional reports. Such documents as theses, dissertations, foreign reports and translations, and non-NRC conference proceedings may be purchased from their sponsoring organization.

Copies of industry codes and standards used in a substantive manner in the NRC regulatory process are maintained at—
          The NRC Technical Library
          Two White Flint North
          11545 Rockville Pike
          Rockville, MD 20852–2738

These standards are available in the library for reference use by the public. Codes and standards are usually copyrighted and may be purchased from the originating organization or, if they are American National Standards, from—
          American National Standards Institute
          11 West 42nd Street
          New York, NY 10036–8002
          www.ansi.org
          212–642–4900

Legally binding regulatory requirements are stated only in laws; NRC regulations; licenses, including technical specifications; or orders, not in NUREG-series publications. The views expressed in contractor-prepared publications in this series are not necessarily those of the NRC.

The NUREG series comprises (1) technical and administrative reports and books prepared by the staff (NUREG–XXXX) or agency contractors (NUREG/CR–XXXX), (2) proceedings of conferences (NUREG/CP–XXXX), (3) reports resulting from international agreements (NUREG/IA–XXXX), (4) brochures (NUREG/BR–XXXX), and (5) compilations of legal decisions and orders of the Commission and Atomic and Safety Licensing Boards and of Directors' decisions under Section 2.206 of NRC's regulations (NUREG–0750).

United States Nuclear Regulatory Commission

Protecting People and the Environment

NUREG/CR-7142
PNNL-21547

Ultrasonic Phased Array Assessment of the Interference Fit and Leak Path of the North Anna Unit 2 Control Rod Drive Mechanism Nozzle 63 with Destructive Validation

Manuscript Completed: July 2012
Date Published: August 2012

Prepared by
S. L. Crawford, A. D. Cinson, P. J. MacFarlan,
B. D. Hanson, R. A. Mathews

Pacific Northwest National Laboratory
P.O. Box 999
Richland, WA 99352

G. Oberson, NRC Project Manager

NRC Job Code N6783

Office of Nuclear Regulatory Research

ABSTRACT

The objective of this investigation was to evaluate the efficacy of ultrasonic testing (UT) for primary water leak path assessments of reactor pressure vessel (RPV) upper head penetrations. Operating reactors have experienced leakage when stress corrosion cracking of nickel-based alloy penetrations allowed primary water into the annulus of the interference fit between the penetration and the low-alloy steel RPV head. In this investigation, UT leak path data were acquired for an Alloy 600 control rod drive mechanism nozzle penetration, referred to as Nozzle 63, which was removed from the North Anna Unit 2 reactor when the RPV head was replaced in 2002. In-service inspection prior to the head replacement indicated that Nozzle 63 had a probable leakage path through the interference fit region. Nozzle 63 was examined using a phased-array UT probe with a 5.0-MHz, eight-element annular array. Immersion data were acquired from the nozzle inner diameter surface. The UT data were interpreted by comparing to responses measured on a mockup penetration with known features. Following acquisition of the UT data, Nozzle 63 was destructively examined to determine if the features identified in the UT examination, including leakage paths and crystalline boric acid deposits, could be visually confirmed. Additional measurements of boric acid deposit thickness and low-alloy steel wastage were made to assess how these factors affect the UT response. The implications of these findings for interpreting UT leak path data are described.

PAPERWORK REDUCTION ACT STATEMENT

PUBLIC PROTECTION NOTIFICATION

FOREWORD

In the previous decade, a number of U.S. pressurized water reactors (PWRs) experienced primary water leakage through cracks in the control rod drive mechanism (CRDM) nozzle penetrations in the upper reactor pressure vessel (RPV) head. At the Davis Besse plant in 2002, such leakage contributed to significant wastage of the low-alloy steel RPV head, leaving only stainless steel cladding at the reactor pressure boundary. The Nuclear Regulatory Commission (NRC) and industry analyses attributed this degradation to primary water stress corrosion cracking (PWSCC). PWSCC affects nickel-based alloys such as those used to fabricate the CRDM nozzle and its associated J-groove weld. The CRDM nozzle has a compression (or interference) fit with bored holes in the RPV head. PWSCC of the J-groove weld allowed primary water leakage into the annulus of the interference fit region, which eventually reached the top of the RPV head at operating pressures. Following the incident at Davis Besse, NRC updated its inspection requirements to mandate that PWR licensees perform a demonstrated surface or volumetric leak path assessment of all J-groove welds in the RPV head. Licensees have proposed to use ultrasonic testing (UT) to satisfy this requirement. In principle, the UT methodology can detect evidence of primary water leakage, such as a flow path in the interference fit between the nozzle and RPV head or crystalline boric acid deposits left in the annulus.

In 2006, the Office of Nuclear Reactor Regulation requested that the Office of Nuclear Regulatory Research evaluate long-term industry solutions for PWSCC of nickel-base primary pressure boundary components. This NUREG/CR documents an evaluation of the UT leak path assessment methodology conducted at Pacific Northwest National Laboratory (PNNL). For this investigation, PNNL used a phased array UT system to acquire leak path data for an Alloy 600 CRDM nozzle, referred to as Nozzle 63, from the North Anna Unit 2 RPV head that was replaced in 2002. Following the acquisition of UT data at PNNL, the nozzle was destructively examined to determine if the features identified in the UT examination could be visually confirmed.

The UT data from PNNL indicated a full leakage path in the interference fit between the penetration tube and the RPV head on the downhill side of Nozzle 63, with suspected boric acid deposits on either side of the flow path. The data also indicated the presence of other partial leakage paths and scattered boric acid deposits between the penetration tube and RPV head. Destructive visual examination confirmed the presence of the leakage path and boric acid deposits in the locations indicated by the UT data. Surface replication of the main leakage path on the RPV head revealed that fabrication machining marks were still visible, indicating minimal loss of material from boric acid corrosion. This suggests that significant wastage of the RPV head material is not necessary for a leak path to develop. The results of this investigation are expected to help staff interpret and evaluate licensees' ultrasonic leak path assessments for upper head penetrations.

Michael J. Case, Director
Division of Engineering
Office of Nuclear Regulatory Research
U.S. Nuclear Regulatory Commission

CONTENTS

FIGURES

TABLES

EXECUTIVE SUMMARY

Research is being conducted for the U.S. Nuclear Regulatory Commission (NRC) at the Pacific Northwest National Laboratory (PNNL) to assess the effectiveness and reliability of advanced nondestructive examination (NDE) methods for the detection and characterization of flaws in nuclear power plant components. One area of concern relates to the nickel-based alloys used in primary pressure boundary components in pressurized water reactors (PWRs). Nickel-based alloys exposed to reactor coolant in PWRs may experience a form of degradation known as primary water stress corrosion cracking (PWSCC). One PWR component that has an operational history of PWSCC is the control rod drive mechanism (CRDM) nozzle. The CRDM nozzles are cylindrical penetrations in the upper reactor pressure vessel (RPV) head that allow for the insertion and removal of control rods. The penetration tube is held in place with an interference fit, and is seal-welded on the underside of the vessel head with a J-groove weld. Cracking in the nozzle or weld metal can allow borated water to leak to the top of the RPV head. Boric acid corrosion of the RPV head, as occurred at Davis Besse, is a concern, as is nozzle ejection in the presence of extensive circumferentially oriented cracking. In response to a number of observations of RPV head leakage at domestic plants, NRC regulations were modified to require PWR licensees to periodically perform a demonstrated surface or volumetric leak path assessment of all J-groove welds in the RPV head.

One plant at which RPV head leakage was observed was North Anna Unit 2. The original construction materials for the CRDM nozzles at North Anna Unit 2 were Alloy 600 base metal and Alloy 82/182 weld metal. During the Fall-2001 refueling outage, coolant leakage was noted near a number of nozzles. In-service inspections showed crack-like indications near the J-groove weld and butter layer in the nozzles and shallow axial cracking on the inner diameter. The leaking welds were repaired by overlaying with Alloy 52/152 material, thought to have higher PWSCC resistance than Alloy 82/182. Subsequent visual examination of the RPV head in the Fall-2002 outage again revealed evidence of leaking nozzles. At that time, the head was replaced and several nozzles, including Nozzle 63 were made available to industry and NRC for study.

PWR licensees have shown interest in using ultrasonic testing (UT) for the required leak path assessments of the upper head penetration. However, the efficacy of UT has not yet been demonstrated by comparing the features detected by UT to the actual features revealed by destructive analysis of a nozzle penetration. Therefore, the purpose of the investigation described in this report was to obtain UT leak path data using Nozzle 63 from North Anna Unit 2, and to validate the findings by opening the nozzle assembly for visual and surface examination of the annulus in the interference fit region.

Before examining Nozzle 63, a phased-array ultrasonic system was calibrated on a mockup CRDM nozzle containing two interference fit regions where low-alloy steel blocks were fit onto an Alloy 600 tube. One interference fit region contained notches with various orientations and dimensions machined onto the surface of the tube and low-alloy steel block, simulating an air gap between the penetration and the RPV head. The probe spot size at the interference fit was modeled at 1.2×1.2 mm (0.04×0.04 in.) at the −6 dB level. The mockup showed that the test system could detect an air gap between the tube and the low-alloy steel block by a high-amplitude ultrasonic signal response compared to areas with no air gap. Ultrasonic data from notches in the low-alloy steel material from one of the mockup interference fit regions showed system resolution at nominally 4 mm (0.16 in.) in both the axial and circumferential directions. Notches as shallow as 0.028 mm (0.0011 in.) were detected as well as notches as narrow as 0.80 mm (0.10 in.) in the circumferential direction. The second interference fit mockup contained regions where crystalline boric acid was placed between the tube and the low-alloy steel block to simulate deposits left from leaking reactor coolant. The mockup showed that the test system could detect the crystalline boric acid by a low-amplitude ultrasonic signal response compared to areas with no boric acid. Taken together, the results from the mockup suggested that the amplitudes of the ultrasonic responses could be used to distinguish three conditions for the interference fit between the penetration and RPV head: 1) nominal interference fit indicated by mid-range amplitude, 2) interference fit with boric acid deposits indicated by low-range amplitude, and 3) leak path or gap based on high amplitude response.

Following the mockup study, ultrasonic data were acquired on Nozzle 63 and clearly showed a variation of responses throughout the annulus region. A region with a high-amplitude response spanned the annulus region at the downhill position of the nozzle. Based on the results from the mockup, this indicated a gap or leakage path between the nozzle penetration and RPV head. Other partial leak paths and scattered boric acid deposits throughout the interference fit were also identified by comparing the ultrasonic signal response to the known response from the mockup. These results compared well with an in-service ultrasonic examination of Nozzle 63 conducted by the licensee prior to replacing the RPV head. After sectioning of the nozzle assembly to reveal the interference fit and visually examining the exposed surfaces, the presence of the primary leak path and boric acid deposits in the areas indicated by the ultrasonic examination were confirmed.

At plant operational temperature and pressure, the gap between the nozzle penetration and the RPV head should open up. If a crack breached the J-groove weld and allowed primary water leakage, the primary water could flow throughout the annulus by the path of least resistance. When the components cool down, the borated water would be trapped between the nozzle penetration and RPV head as the gap closes. The boric acid may be compacted to some extent based on the size of the gap at room temperature. The boric acid may also cause corrosion on the surface of the RPV head material in the annulus leaving corrosion deposits. The thickness and compactness of the boric acid deposits and the extent of RPV head surface corrosion may affect the ultrasonic signal response.

To evaluate the correlation between the surface deposit thickness and ultrasonic signal response, the deposit thicknesses on the RPV annulus surface were measured using an eddy current coating thickness gage. Regions corresponding to the leak path or that visually appeared as bare metal had deposit thicknesses of 16 microns (0.63 mils) or less, with ultrasonic responses greater than 65% of the full-screen height. Boric acid apparently did not deposit in the primary leak path between the nozzle penetration and RPV head due to the flow of borated water through the area, leaving a gap instead. On both sides of the primary leak path, the ultrasonic responses quickly drop off and the surface deposit or corrosion layer thickness increases. This inverse relationship with a sharp transition at the edge of the leak path defines one of the leak path characteristics. Another characteristic of a leak path is that it fully traverses the interference fit zone. The area of the leakage path was further analyzed by surface replication. The replica surfaces were imaged with a stereomicroscope. Two small areas with minimal corrosion were found but most of the leak path showed only minor evidence of corrosion product streaking with clearly visible fabrication machining marks, indicating little or no loss of material.

Outside of the primary leak path, the surface deposit or corrosion layer thickness within the span of the interference fit, as measured by the eddy current probe, is generally less than 75 microns (3.0 mils), and the ultrasonic response has low amplitude (less than 50% full-screen height), indicating good energy transmission between the nozzle penetration and RPV head. This suggests that the deposits within the span of the interference fit are relatively compact or dense because the gap between the nozzle penetration and RPV head is narrow at room temperature. Conversely, the boric acid deposits above or below the interference fit are thicker, in the range of 130–190 micron (5.12–7.48 mils). The ultrasonic responses are higher (greater than 50% full-screen height) indicating relatively poor energy transmission between the nozzle penetration and RPV head. This suggests that the deposits above or below the span of the interference fit are less compact or dense because the gap between the nozzle penetration and RPV head is relatively wide at room temperature.

This investigation clearly demonstrated the efficacy of an encoded phased-array ultrasonic evaluation of the interference fit in Nozzle 63. Visual examinations of the exposed surfaces in the interference fit region confirmed the ultrasonic findings and deposit layer thickness measurements on the RPV head provided further insights.

ACKNOWLEDGMENTS

The work reported here was sponsored by the U.S. Nuclear Regulatory Commission (NRC) and conducted under NRC Job Code Number N6783. Greg Oberson is the NRC project manager. The Pacific Northwest National Laboratory (PNNL) would like to thank Dr. Oberson, Mr. Darrell Dunn, and Mr. Jay Collins for their guidance throughout the course of this effort.

Dr. Stephen Cumblidge is acknowledged and thanked for initiating this project.

The authors acknowledge and thank J. W. Hyres, et al. at Babcock & Wilcox Technical Services Group in Lynchburg, Virginia, for cutting the nozzle assembly, for acquiring additional measurements on the head material, and for excellent photography and documentation of their work. A special thanks to Jim for his flexibility and willingness to try new approaches to obtain requested data is due.

The authors acknowledge and thank John P. (Jack) Lareau from WesDyne International for providing information on in-service inspections (ISI) and nozzle fabrication as well as data from an ISI on Nozzle 63.

The authors also thank Kay Hass, PNNL, for her diligence and patience in editing and preparing the manuscript.

PNNL is operated by Battelle for the U.S. Department of Energy under Contract DE-AC05-76RL01830.

ACRONYMS AND ABBREVIATIONS

ASME	American Society of Mechanical Engineers
BA	boric acid
B&W	Babcock and Wilcox Technical Services Group
CFR	Code of Federal Regulations
CRDM	control rod drive mechanism
dB	decibels
EC	eddy current
EDM	electric discharge machining
EPRI	Electric Power Research Institute
FSH	full-screen height
I Fit	interference fit
ID	inner diameter
IR	infrared
ISI	in-service inspection
LN	liquid nitrogen
LWR	light water reactor
NDE	nondestructive examination
NRC	U.S. Nuclear Regulatory Commission
OD	outer diameter
PA	phased array
PA-UT	phased-array ultrasonic testing
PE	pulse-echo
PNNL	Pacific Northwest National Laboratory
PWR	pressurized water reactor
PWSCC	primary water stress corrosion cracking
RMSE	root mean squared error
RPL	Radiochemical Processing Laboratory
RPV	reactor pressure vessel
RT	room temperature
TOFD	time-of-flight diffraction
UT	ultrasonic testing

1 INTRODUCTION

Research is being conducted for the U.S. Nuclear Regulatory Commission (NRC) at the Pacific Northwest National Laboratory (PNNL) to assess the effectiveness and reliability of advanced nondestructive examination (NDE) methods for the detection and characterization of flaws in nuclear power plant components. One area of concern relates to the nickel-based alloys used in primary pressure boundary components in pressurized water reactors (PWRs). Nickel-based alloys exposed to reactor coolant in PWRs may experience a form of degradation known as primary water stress corrosion cracking (PWSCC). One PWR component that has an operational history of PWSCC is the control rod drive mechanism (CRDM) nozzle. As shown in Figure 1.1, the CRDM nozzles are cylindrical penetrations in the upper reactor pressure vessel (RPV) head that allow for the insertion and removal of control rods. The penetration tube is held in place with an interference fit, represented as the area labeled 'shrink fit zone' between the two horizontal dashed lines in the figure, and is seal-welded to the underside of the vessel head with a J-groove weld. Counter bore regions are not designed to be compression-fit zones between the nozzle and RPV head and are shown at exaggerated scale in the figure. Most CRDM nozzles originally placed into service in PWRs were fabricated from the nickel-based alloy referred to as Alloy 600, along with the Alloy 82 and 182 J-groove weld metals. PWSCC of a CRDM nozzle in a PWR was first identified in the Bugey Unit 3 plant in France during an over-pressurization test in 1991 (Economou et al. 1994). The crack initiated in the Alloy 600 base metal and propagated into the Alloy 182 weld metal. In late 2000 and early 2001, reactor coolant leakage to the RPV head from axial through-wall cracks in CRDM nozzles was identified at Arkansas Nuclear One Unit 1 and Oconee Unit 1 (Grimmel 2005). Follow-up inspections at Oconee Units 2 and 3 in 2001 identified axially and circumferentially oriented cracks. The circumferentially oriented cracks were of particular concern because of the possibility of nozzle ejection.

Leakage of borated water to the RPV head may occur as cracks initiate on the J-groove weld surface, propagate through the weld to the triple point, and allow water into the annulus region between the nozzle outer diameter (OD) and the RPV head. The triple point is diagrammed in Figure 1.2 and is the point at which the RPV head, weld buttering, and Alloy 600 CRDM tube meet. Once the boundary formed by an intact J-groove weld is compromised, there is the potential for a leakage path through the interference fit allowing reactor coolant to reach the outer surface of the RPV. The coolant can flash to steam, leaving boric acid deposits on the head and in the interference fit region around the leakage path. Additionally, a steam-cut leakage path through the interference fit and annulus may also be produced at operating temperature and pressure in a plant when a gap in the low-alloy steel RPV head at the uphill and downhill positions opens due to material expansions.

In response to the discovery of the CRDM cracks at Oconee Unit 3, in August 2001, the NRC issued Bulletin 2001-01, "Circumferential Cracking of Reactor Pressure Vessel Head Penetration Nozzles." PWR licensees were directed to evaluate the susceptibility of head penetration nozzles to PWSCC and to provide inspection plans to detect potential cracking. Thereafter, CRDM cracking was identified at additional PWRs including Davis Besse (Bennetch et al. 2002) and North Anna Unit 2 (NRC 2002). At Davis Besse, reactor coolant leakage led to significant wastage of the low-alloy steel in a portion of the RPV head, leaving only a thin layer of stainless steel cladding at the pressure boundary. In response to the occurrences of RPV head leakage, in 2004 NRC issued EA-03-009 for PWR licensees requiring additional periodic inspections and evaluation of boric acid deposits as they pertain to the reasonable assurance of plant operational safety. The requirements of EA-03-009 were superseded by the adoption of American Society of Mechanical Engineering (ASME) Section XI Code Case N-729-1 by rulemaking in Title 10 of the Code of Federal Regulations (10 CFR) Part 50.55(a)(g)(6)(ii)(D)(1). As a condition in 10 CFR 50.55(a)(g)(6)(ii)(D)(3), licensees are directed to perform a demonstrated surface or volumetric leak path assessment of all J-groove welds in the RPV head.

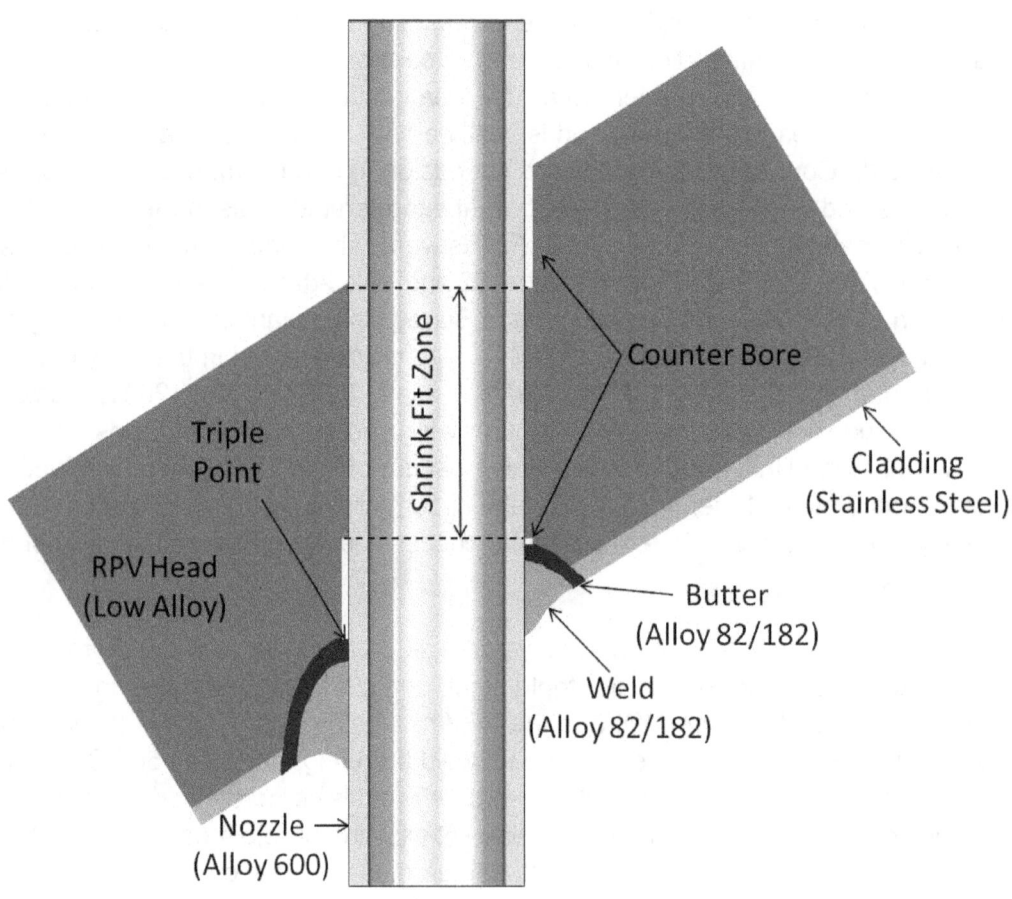

Figure 1.1 CRDM J-groove Weld Schematic

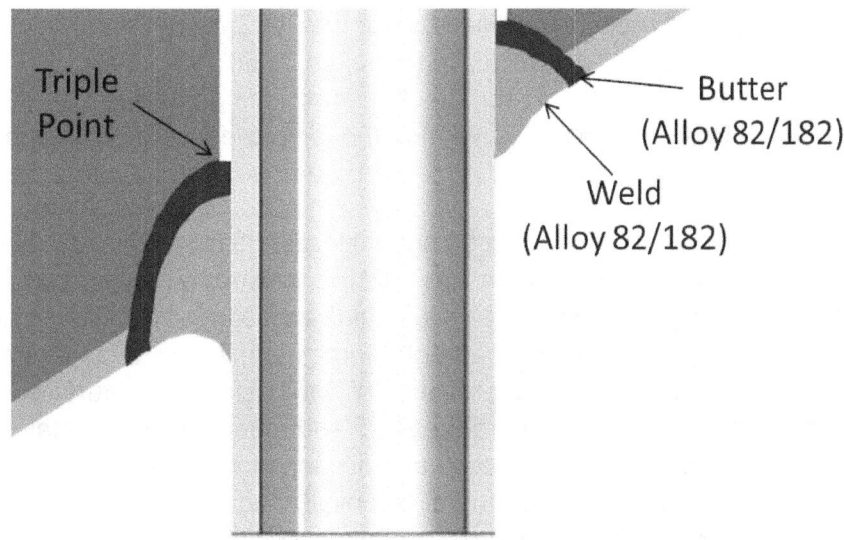

Figure 1.2   The Triple Point in the Assembly Where the Alloy 600 Nozzle, RPV Head, and Weld Buttering Material Meet

A leak path assessment involves the use of a NDE technique, such as ultrasonic testing (UT), to determine whether a flow path exists through the interference fit that would allow reactor coolant to access the outside of the RPV head. An effective ultrasonic examination of the interference fit should detect a leak path and may additionally detect corrosion or loss of material as well as the presence of boric acid in the annulus region. Industry groups, such as the Electric Power Research Institute (EPRI), are participating in programs to generically demonstrate the viability of the ultrasonic leak path assessment methodology. However, the efficacy of UT has not yet been investigated by comparing the features detected by UT to the actual features revealed by destructive analysis. Therefore, the purpose of the investigation described in this report was to obtain UT leak path data from a removed-from-service CRDM nozzle, and to validate the findings by opening the nozzle assembly to visually examine the annulus surfaces.

The subject nozzle for this investigation, designated Nozzle 63 by the licensee, was removed from the North Anna Unit 2 plant. North Anna Unit 2 is a three-loop Westinghouse PWR that was placed into service in 1980. The materials of construction for the original CRDM nozzles were Alloy 600 base metal and Alloy 82/182 weld metal. Visual inspection of the outer surface of the North Anna Unit 2 RPV head during the Fall-2001 refueling outage indicated reactor coolant leakage in the proximity of penetrations 51, 62, and 63 as evinced by the presence of boric acid crystals (EPRI 2005). In-service ultrasonic examination of the nozzles showed crack-like indications near the J-groove weld/butter layer in the nozzles and shallow axial cracking on the inner diameter. The leaking welds in these nozzles were repaired using a temper bead repair technique with nickel-based Alloys 52 and 152, which are thought to have higher PWSCC resistance than Alloy 182. Subsequent visual examination of the RPV head during the Fall-2002 outage revealed six CRDM nozzles that were suspected of leaking and 21 that were masked to the extent that their status could not be determined. Eddy current and ultrasonic examinations showed numerous axial and circumferential indications in the nozzles, including those repaired during the previous outage. Given the extensive degradation of the RPV head, the utility made the decision to replace the head during the 2002 outage and make certain penetrations available to industry and NRC for further examination and study. EPRI took possession of six CRDM nozzles from the removed head including nozzles 10, 31, 51, 54, 59, and 63, which were transferred to PNNL. Several of these nozzles were subsequently studied by EPRI and the NRC including nozzles 10, 31, 54, and 59 (EPRI 2006; Cumblidge et al. 2009).

The subject of this report is a leak path assessment of Nozzle 63 from North Anna Unit 2. This nozzle is of additional interest because of the Alloy 52/152 weld repair during the Fall-2001 outage. The purpose of this investigation is to evaluate the efficacy of the ultrasonic leak path assessment methodology by determining whether features identified by a UT examination of the nozzle, including leak paths, voids, and the presence of boric acid in the interference fit, are confirmed by destructive examination. The radiological and mechanical steps taken to configure the nozzle for the ultrasonic evaluation are discussed in Section 2 of this report. Section 3 presents technical information on the ultrasonic transducer, the system electronics, and mechanical scanner. Based on these specifications, the UT process is assumed to be equivalent to or better than that used in industry examinations. A calibration mockup specimen is described in Section 4. Ultrasonic data on the mockup specimens are presented and system resolution and flaw detection capabilities are discussed. The ultrasonic evaluation of Nozzle 63 and corresponding results are presented in Section 5. Section 6 documents the cutting activities on the nozzle assembly to reveal the interference fit. Initial views of the annulus region are shown. Section 7 compares the ultrasonic findings to the visual evaluation of the interference fit. Section 8 contains additional measurements on the RPV head including boric acid thickness measurements in the annulus and Microset replication of the primary leak path region. Lastly, a summary of the findings are presented in Section 9. Section 10 gives references cited in this report.

2 NOZZLE PREPARATION

When received at the Pacific Northwest National Laboratory (PNNL), the control rod drive mechanism (CRDM) Nozzle 63 from the North Anna Unit 2 reactor consisted of a flame-cut section of the upper reactor pressure vessel (RPV) head and a full-length Alloy 600 penetration tube, as shown in Figure 2.1. The CRDM was removed from its storage box and a radiologic survey performed. The flame-cut edges of the RPV head around the penetration were then painted with two coats (the first yellow, the second red) of a flexible, air-dried plastic coating from Plasti Dip to reduce the risk of workers being cut while handling the CRDM (Figure 2.2). An expandable 7.62-cm (3-in.) plug was inserted in the wetted side of the penetration tube so the tube could be filled with water. In retrospect, it would have been better to cut the nozzle prior to inserting the plug. This may have reduced some of the debris that first coated the ultrasonic scanner (see Section 5.2). With the plug inserted, the nozzle was wrapped in plastic (Figure 2.3) and bagged to contain contamination during the nozzle cutting. The nozzle was then secured on a wheeled cart that was modified to allow the penetration tube to be kept vertical during the testing, as seen in Figure 2.4.

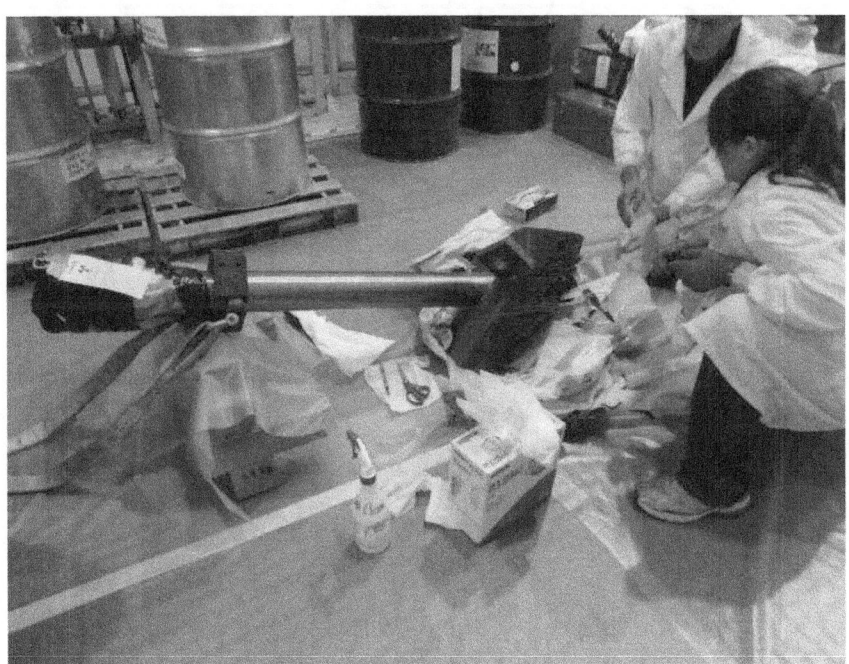

Figure 2.1 As-Received Condition of Nozzle 63

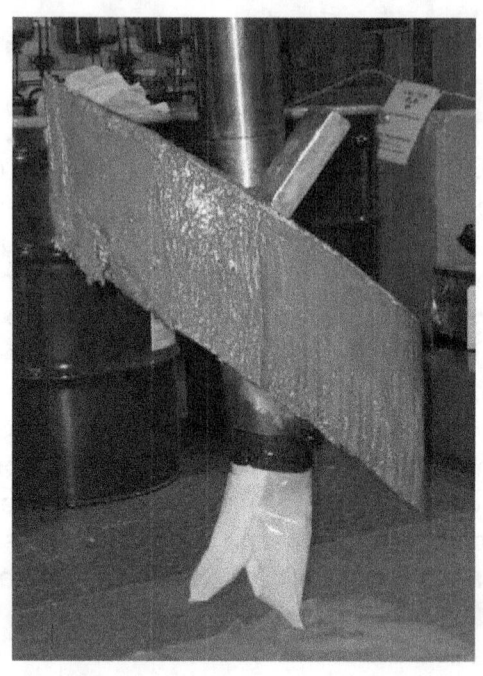

Figure 2.2 Nozzle 63 with Painted Edges

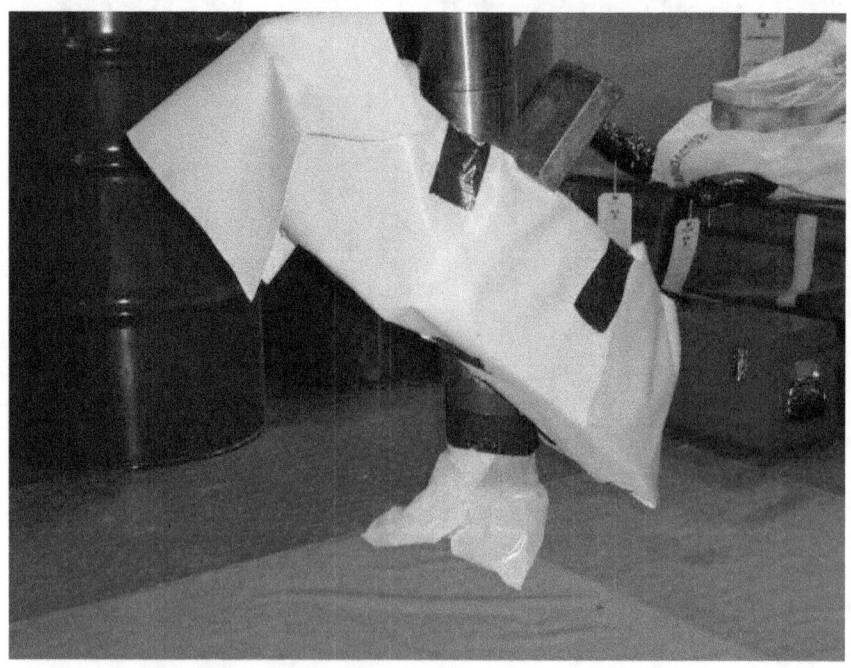

Figure 2.3 Nozzle 63 Wrapped in Plastic for Contamination Control

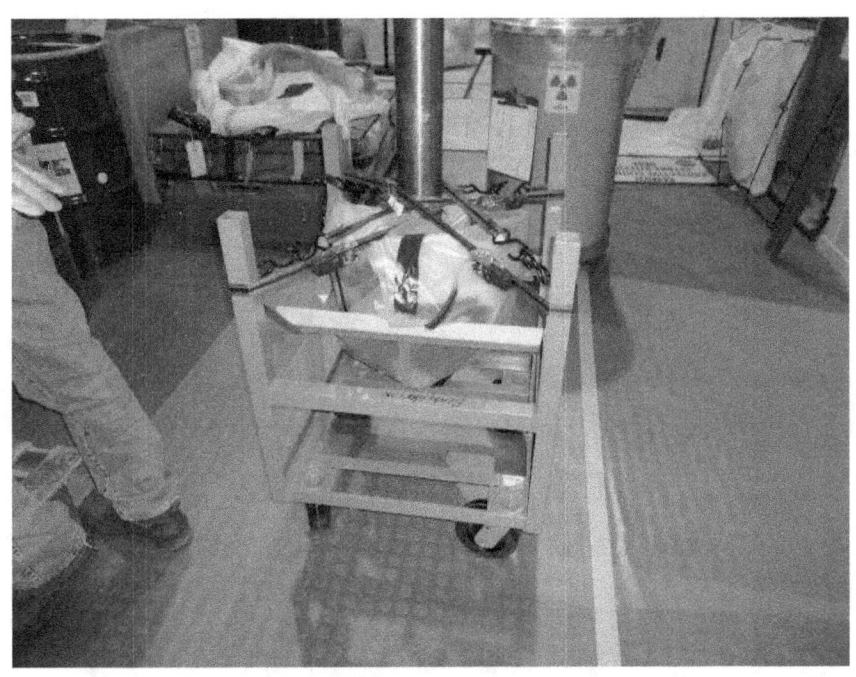

Figure 2.4 Nozzle 63 on Modified Cart

In order to better fit in a glovebag for contamination-control purposes and to facilitate connection of the scanner, approximately 61 cm (2 ft) of the penetration tube had to be removed. A catch pan was first fitted around the penetration tube below where the cut was to be made, approximately 30 cm (12 in.) from what was the top side of the RPV head. A hydraulic rotary pipe-cutting tool was then fitted around the penetration tube as shown in Figure 2.5. The penetration tube and catch tray were wrapped in plastic to provide contamination control. The hydraulic cutting tool was connected and the first attempt at cutting the tube was made (Figure 2.6). A hard spot was encountered within the nozzle and two cutting heads broke before the decision was made to attempt a new cut approximately 2.54 cm (1 in.) above the previous cut attempt. Cutting proceeded without any other issues in this new location. The removed section of the penetration tube was stored in a 208-L (55-gal.) drum.

After cutting, the CRDM was completely wrapped in plastic to prevent the spread of contamination. The CRDM was then transported to the PNNL's Radiochemical Processing Laboratory (RPL/33) where a containment glovebag had been assembled to house the nozzle. The cart with the nozzle was wheeled into the glovebag as shown in Figure 2.7. Once the nozzle was properly positioned, the two-axis scanner with attached ultrasonic phased-array probe was lowered through an upper access port and centered onto the penetration tube. The scanner was secured to the penetration tube using three set screws. The main door and the upper access port were sealed and the control cables secured to the upper frame. Approximately 1.5 L (0.39 gal.) of distilled water was added to the penetration tube. The final configuration of the scanner in the glovebag is shown in Figure 2.8.

Figure 2.5 Nozzle 63 with Hydraulic Rotary Pipe Cutting Tool

Figure 2.6 Cutting of Nozzle 63 Penetration Tube

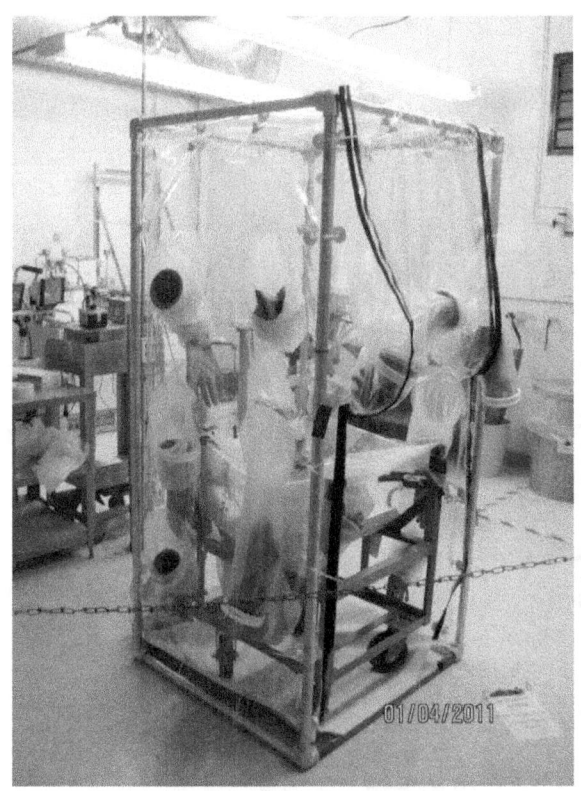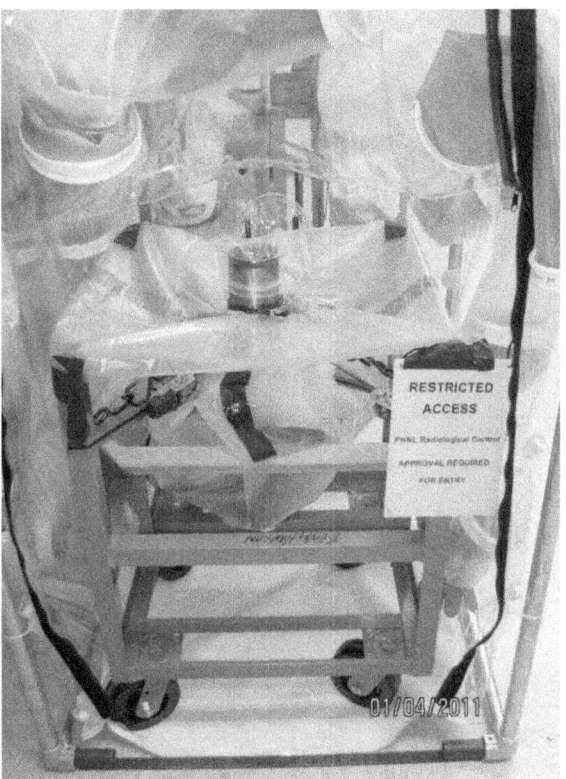

RESTRICTED
ACCESS

PNNL Radiological Control
APPROVAL REQUIRED
FOR ENTRY

Figure 2.7 CRDM Installed in Containment Glovebag

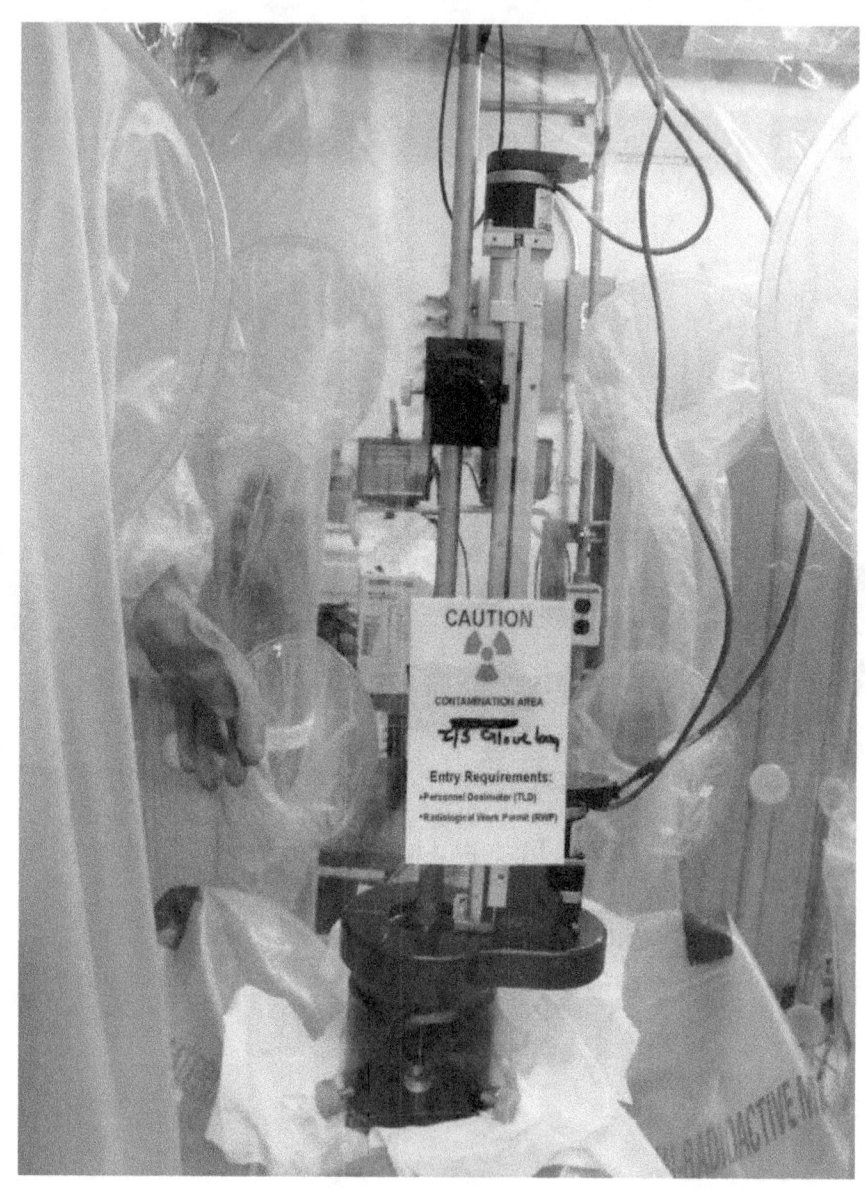

Figure 2.8 Final Configuration of CRDM Nozzle 63 for Examination

3 ULTRASONIC TESTING EQUIPMENT FOR NOZZLE 63 EXAMINATION

The nondestructive leak path assessment of Nozzle 63 was performed at the Pacific Northwest National Laboratory (PNNL) with an ultrasonic phased-array (PA) probe. A PA probe has multiple individual elements that are electronically fired at prescribed time delays to form coherent and focused beams at specified depths and angles in the material under examination. The selected PA ultrasonic equipment provided high-resolution, spatially encoded ultrasonic data for analysis of leak paths present in the Nozzle 63 interference fit region. An advantage of using PA equipment over conventional single-element equipment (typically used by industry) is that multiple foci can be employed during a single scan sequence on the specimen. As such, the equipment used for this investigation was selected because it is similar to or better than equipment used by industry for in-service inspections of upper reactor pressure vessel (RPV) nozzle penetrations in pressurized water reactors (PWRs). A detailed description and justification of the equipment used is provided in this section.

3.1 Phased-Array Electronics

Ultrasonic data acquisition for Nozzle 63 was accomplished using the ZETEC Tomoscan III PA system to control the PA probe employed in this study. This commercially available system was equipped to accommodate a maximum of 64 channels of data from PA probes and was operated with UltraVision 1.2R4 software. Its electronics can drive probes in the 0.7–20 MHz frequency range. The system is capable of accepting multiple axis positional information from external encoders to map ultrasonic data to spatial location on a specimen. The data acquisition system is shown in Figure 3.1.

3.2 Phased-Array Probe and Software Simulations

Nozzle 63 was examined with a pulse-echo (PE) longitudinal-wave immersion PA probe with a center frequency of 5 MHz, as shown in Figure 3.2. At 5 MHz, the probe is able to penetrate through the Alloy 600 tube to the interference fit and maintain a good resolution for flaw detection as the wavelength is 1.1 mm (0.043 in.) in the tube material. This was one of the frequencies used by industry as well. The PA probe was designed in a 1-D annular configuration using eight elements for a normal beam or zero-degree inspection. The probe contained elements in a Fresnel radius pattern starting with a radius of 3 mm (0.12 in.) up to the final element radius of 9.72 mm (0.38 in.). Thus, the total aperture was 296.81 mm^2 (0.46 in.^{2}). As characterized by the manufacturer, Imasonic, the probe exhibited an overall 71% bandwidth at −6 decibels (dB) with all eight elements and an overall central frequency of 5.4 MHz. This design was chosen for enhanced depth focusing capabilities. Its beam-forming feature showed ideal insonification of the interference fit region of interest as well as the ability to propagate a coherent ultrasonic beam deep into the weld region. Figure 3.3 shows the probe attached to the scanning arm.

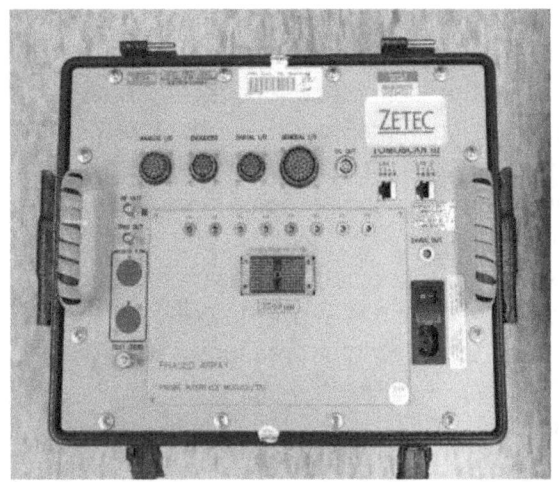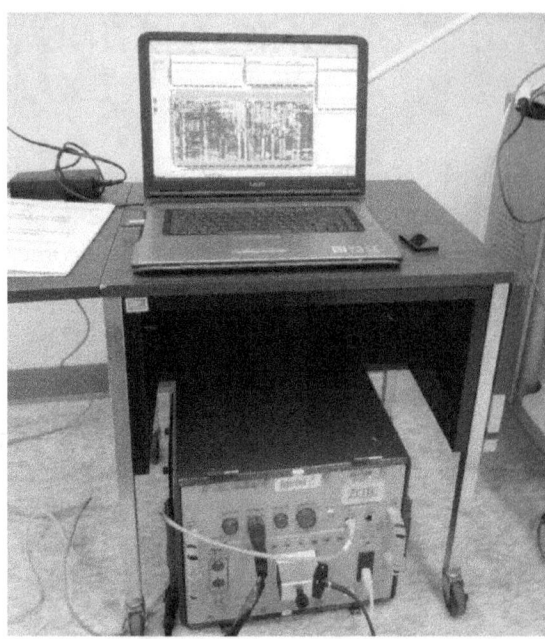

Figure 3.1   Data Acquisition System and Laboratory Workstation. Left: Tomoscan III phased-array data acquisition system. Right: Laboratory workstation/laptop computer for both data acquisition and data analysis, with the Tomoscan III system below.

Before the PA probe was used for the examination, a set of focal laws was produced to control the firing of individual elements. The focal laws were inputs to the UltraVision control software, which determined specific elements to excite at specific times to allow for proper beam-forming in the material. The focal laws may also contain details about insonification angles, the focal depth of the sound field, the delays associated with the wedge and electronics, and the orientation of the probe. For this investigation, a software package contained in the UltraVision software program suite, known as the ZETEC Advanced Focal Law Calculator 1.2R4, was used to produce the focal laws. The software program generated focal laws and simulated the ultrasonic field produced by the probe when using the generated laws. The user entered the physical information about the PA probe and wedge into the program, including the number and size of probe elements, and the wedge angle and size. After the desired angles and focal distances were entered, the software generated the needed delays for each element to produce the desired beam steering and focusing in the material. The software beam simulation produced a simple ray-tracing image of the probe, wedge, and material under evaluation, as well as a density mapping of the modeled sound field. The sound field mapping enabled the user to see how well the sound field was formed with the given input parameters. The probe was also evaluated for the generation of grating lobes that may be detrimental to the examination. It should be noted that the software simulation was performed using an isotropic material assumption; namely, that the velocity of sound is maintained throughout any angle for a particular wave mode. The simulations enabled the user to estimate sound field parameters and transducer performance to optimize array design and focal law development.

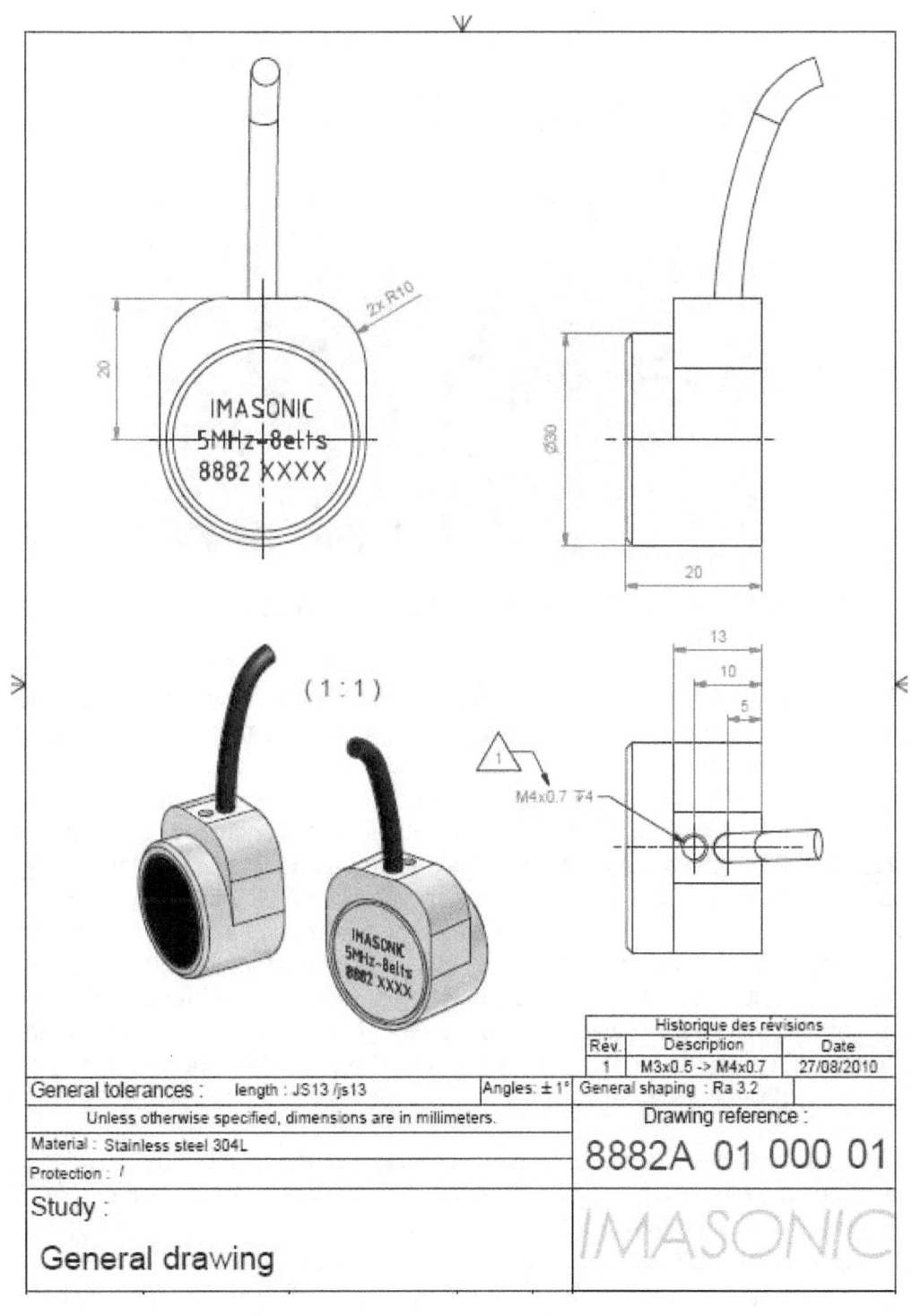

Figure 3.2 5.0-MHz Phased-Array Probe

Figure 3.3 Annular Phased-Array Probe Attached to Scanner Arm

Typical control rod drive mechanism (CRDM) nozzles made from Alloy 600 have a tube wall thickness on the order of 15–17 mm (0.59–0.67 in.). Because the targeted area of interest in this study was the interference fit in the annulus between the outer diameter (OD) of the nozzle and the RPV head, it was important to design a PA probe capable of depth focusing into this region. Prior to probe fabrication, sound field simulations were conducted using the Phased Array Calculator 1.2R4 software program and the design parameters to simulate a projected sound field into an isotropic material with acoustic properties of Alloy 600. Figure 3.4 shows a side view representation of the focal laws generated on the left and a sound field simulation on the right for a target depth focus of 15 mm (0.59 in.). The gray-green regions at bottom represent the Alloy 600 material and the dark blue regions represent water as labeled. In this immersion scanning setup, water was used as the 'wedge' material. The red horizontal line at 15.1 mm (0.59 in.) represents the target focal region. The simulation showed a favorable sound field density at the desired focal depth.

The simulations viewed from the top or C-scan view gave information on the overall spot size in the scan and index axes of the formed beam at a particular depth. As seen in Figure 3.5, the predicted −6 dB (50%) and −3 dB (70.7%) spot sizes for the PA probe focused at a depth of 15 mm (0.59 in.) in Alloy 600 material were 1.2 × 1.2 mm (0.047 × 0.047 in.) and 1.0 × 1.0 mm (0.04 × 0.04 in.), respectively. Additional sound field simulations were modeled at 1.0 mm (0.04 in.), corresponding to the tube inner diameter (ID) and 30 mm (1.18 in.), which is 15 mm (0.59 in.) into the J-groove weld region. The −6 dB spot size for depth foci of 1 and 30 mm (0.04 and 0.59 in.) were 0.6 × 0.6 mm (0.024 × 0.024 in.) and 2.0 × 2.0 mm (0.079 × 0.079 in.), respectively. The −3 dB spot sizes were 0.4 × 0.4 mm (0.016 × 0.016 in.) and 1.4 × 1.4 mm (0.055 × 0.055 in.), respectively.

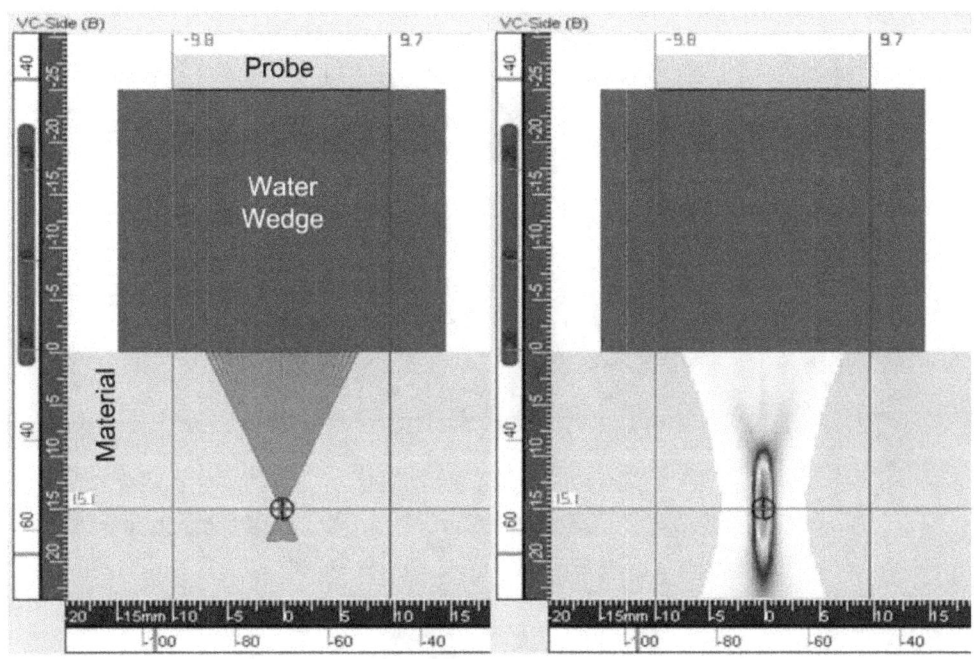

Figure 3.4   Side View—Left: Law Formation. Right: Sound Field Simulation for a Depth Focus of 15 mm (0.59 in.). Blue and red lines are measurement cursors from the origin defined as center of probe in X,Y and material face in Z.

Typical conventional probes used by one in-service inspection vendor had frequencies of 5 and 2.25 MHz with nominal single element diameters of 6.35 mm (0.25 in.) or 4.76 mm (0.19 in.) respectively. The nominal near field position or position at which the probe starts to focus, for a 5-MHz, 4.76-mm (0.19-in.) diameter transducer is 5.0 mm (0.20 in.). The estimated ultrasonic beam spread occurs over an angle of 14 degrees through the 15-mm (0.59-in.) wall thickness of the Alloy 600 tube resulting in a theoretical spot size of 3.8 × 3.8 mm (0.15 × 0.15 in.) at the interference fit region. For reference, the modeled spot size of the 5-MHz, 8-element annular PA probe at the interference fit region is 1.2 × 1.2 mm (0.047 × 0.047 in.). A reduced spot size will be more sensitive to small flaws and provide improved resolution and accuracy in defect sizing.

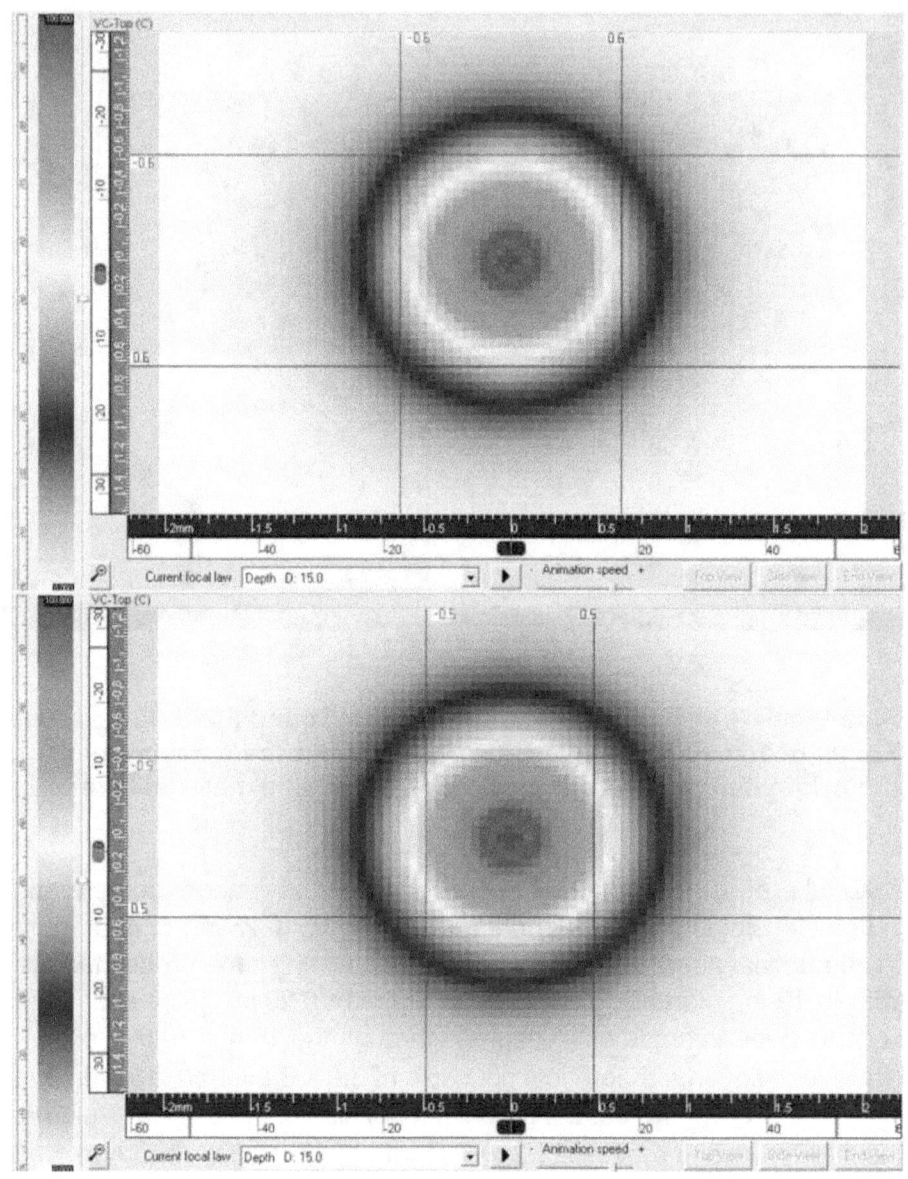

Figure 3.5   C-Scan View at a Depth Focus of 15 mm (0.59 in.). Top: −6 dB spot size. Bottom: −3 dB spot size. Blue and red lines are measurement cursors from the origin defined as center of probe in X,Y. The cursors are positioned at the -6 and -3 dB locations.

3.3 Scanner

The theta-Z scanning apparatus for examining the annulus region of the nozzle from the ID of the penetration tube was constructed by Brockman Precision Machine and Design located in Kennewick, Washington. The ID scanner was designed for inner-surface scanning with ultrasonic probes but could be adapted and used with other sensor technologies. The scanner system was built to attach directly and securely onto the nozzle, centered by three set screws spaced evenly around the collar of the scanner. Figure 3.6 shows photographs of the scanner sitting on a nozzle mockup specimen. The scanner had a linear, Z or vertical axis for movement along the length of the nozzle and a rotational or theta axis for rotation around the nozzle. Motion of the scanner was controlled by two pulse-counter or stepper motors. Optical eye shaft encoders with a sensitivity of 2500 counts per revolution were attached to each motor. The calibrated positional information attained via the slave encoders was routed directly into the ultrasonic system and correlated with the ultrasonic testing (UT) data. The maximum range of motion along the nozzle length was 457.2 mm (18 in.). The rotational motion was continuous with no fixed limits, but was practically constrained to approximately 1.5 revolutions by the cables attached to the motor drivers, encoders, and the PA probe.

The scanner system was controlled using a custom-designed software program interfaced with a pulse-counter motor control system. A menu in the program allowed the user to 'jog' the scanner to a desired position. This feature was useful for setup, mapping the desired scan bounds as well as calibrating the UT signal response at certain locations. The customizable scanning sequence menu allowed the user to specify the scan and index range and resolution settings. Additionally, speed settings were tailored to acquire data with consistency and within the UT data acquisition system limits.

Prior to scanning, the nozzle was oriented vertically, plugged with a water-tight seal in the bottom end, and then filled with distilled water. In immersion scanning, water serves as both the wedge material and the ultrasonic couplant material. The water was given 24 hours or more to degas/de-bubble. Next, the ID of the nozzle specimen was gently brushed to remove bubbles that formed and attached to the ID wall region. Because air bubbles have a strong ultrasonic impedance mismatch to water or steel, it was important to remove them from the ID surface prior to scanning to minimize reflection or distortion of the ultrasonic energy.

The scanner was lowered onto the top of the nozzle specimen, centered, and secured by uniformly tightening the three set screws in the collar. Centering the scanner apparatus allowed the transducer arm to be positioned at the center of the nozzle tube so that a constant sound path was maintained during a circumferential scan sweep to reduce signal walk. The phased-array probe was then affixed to the 762-mm (30-in.) scanner arm using an M4 threaded rod running directly into the transducer housing and attached to the vertical axis via a set screw, as shown in Figure 3.7. The transducer face was orientated such that the ultrasonic beam was propagating radially outward towards the annulus or weld region. Using a set screw to hold the scanner arm enabled manual positioning of the probe in the vertical axis for increased versatility.

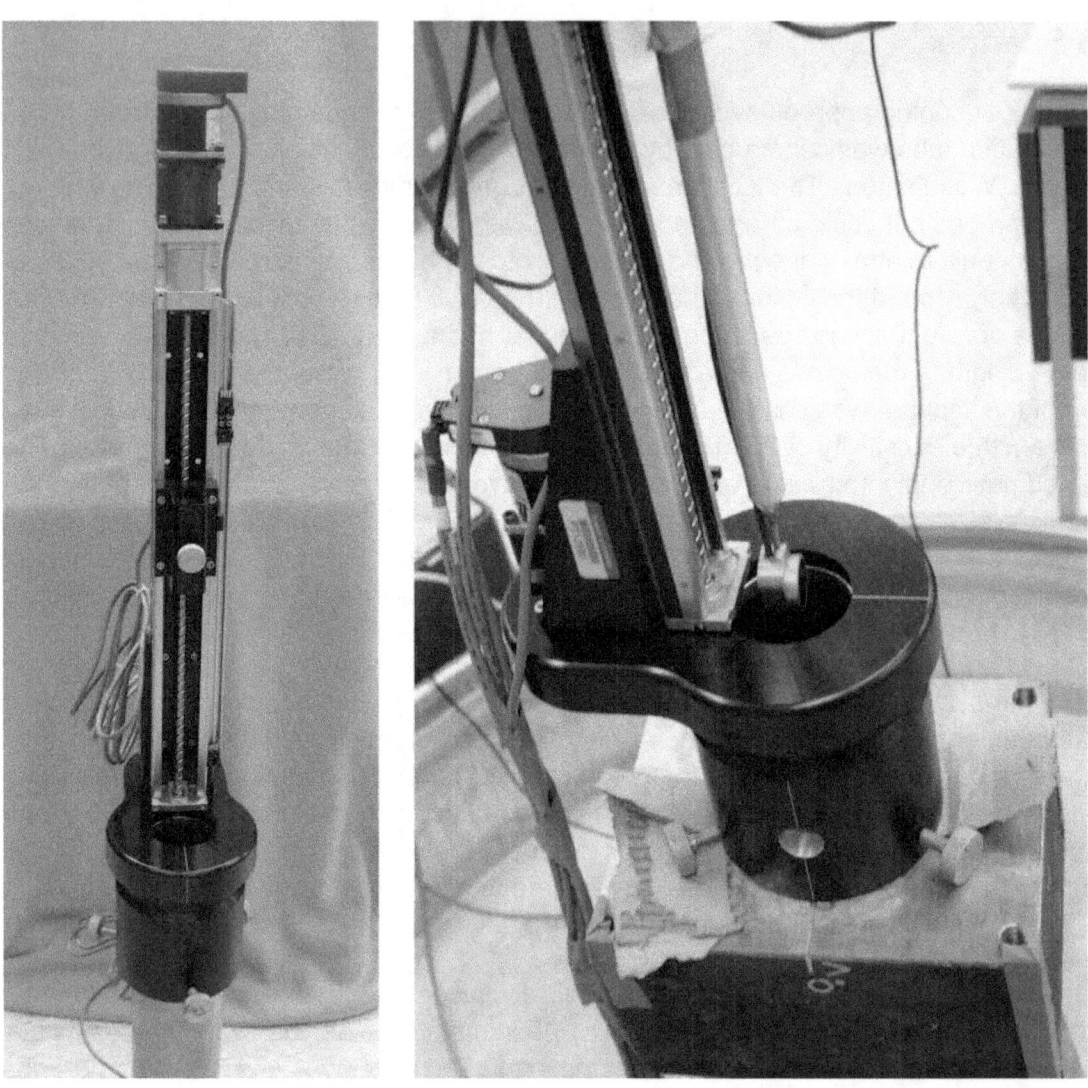

Figure 3.6   Scanner on Mockup Nozzle Specimen. Left: Scanner alone. Right: Scanner with PA probe attached sitting on the calibration mockup specimen.

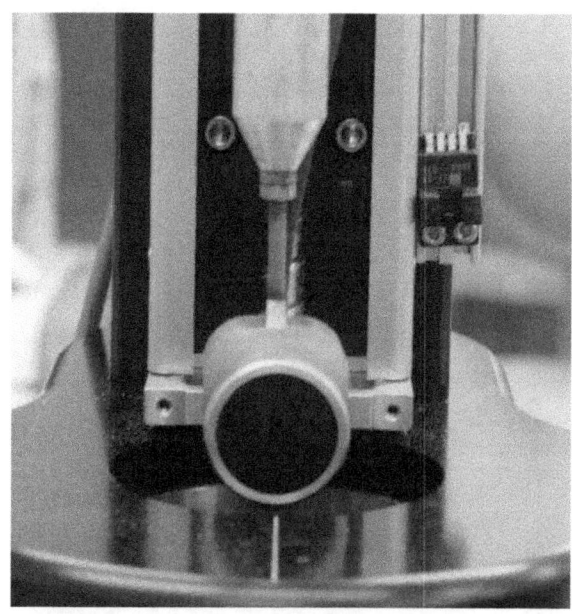

Figure 3.7 Transducer Attachment

The scanning sequence used the rotational or circumferential direction as the scan axis and the vertical direction as the index axis. The positive scan direction was established to be counter clockwise and the positive index was defined as vertically upwards. Positional resolutions were set to 0.25 degrees in the scan and 0.5 mm (0.02 in.) in the index directions for scanning the calibration mockup specimen. For output file size management, Nozzle 63 scanning protocol used 0.5 degrees by 0.5-mm (0.02-in.) resolutions in the scan and index directions, respectively. Figure 3.8 shows a detailed scanning setup schematic on a CRDM nozzle assembly.

For radiological control, a custom glovebag (details discussed in Section 2) was constructed around Nozzle 63 to reduce radiation contamination to persons or equipment. Setup in the glovebag required modifications to the glovebag so that scanner and phased-array cables and equipment could be passed in while maintaining connection to vital equipment such as the phased-array electronics and motor control units. Figure 3.9 depicts the scanner system fully assembled in the protective glovebag environment.

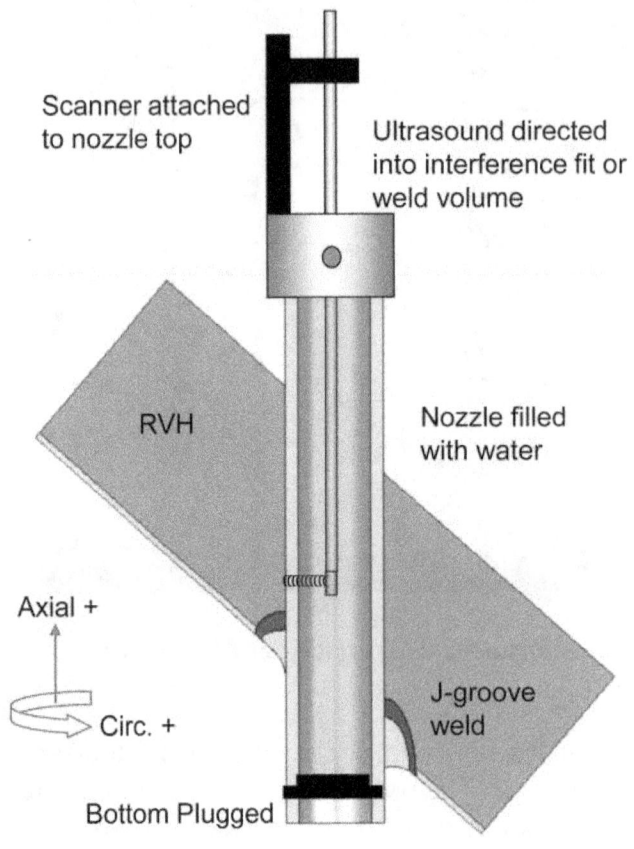

Scanner attached
to nozzle top

Ultrasound directed
into interference fit or
weld volume

RVH

Nozzle filled
with water

Axial +

Circ. +

J-groove
weld

Bottom Plugged

Figure 3.8 Scanning Setup/Orientation Schematic

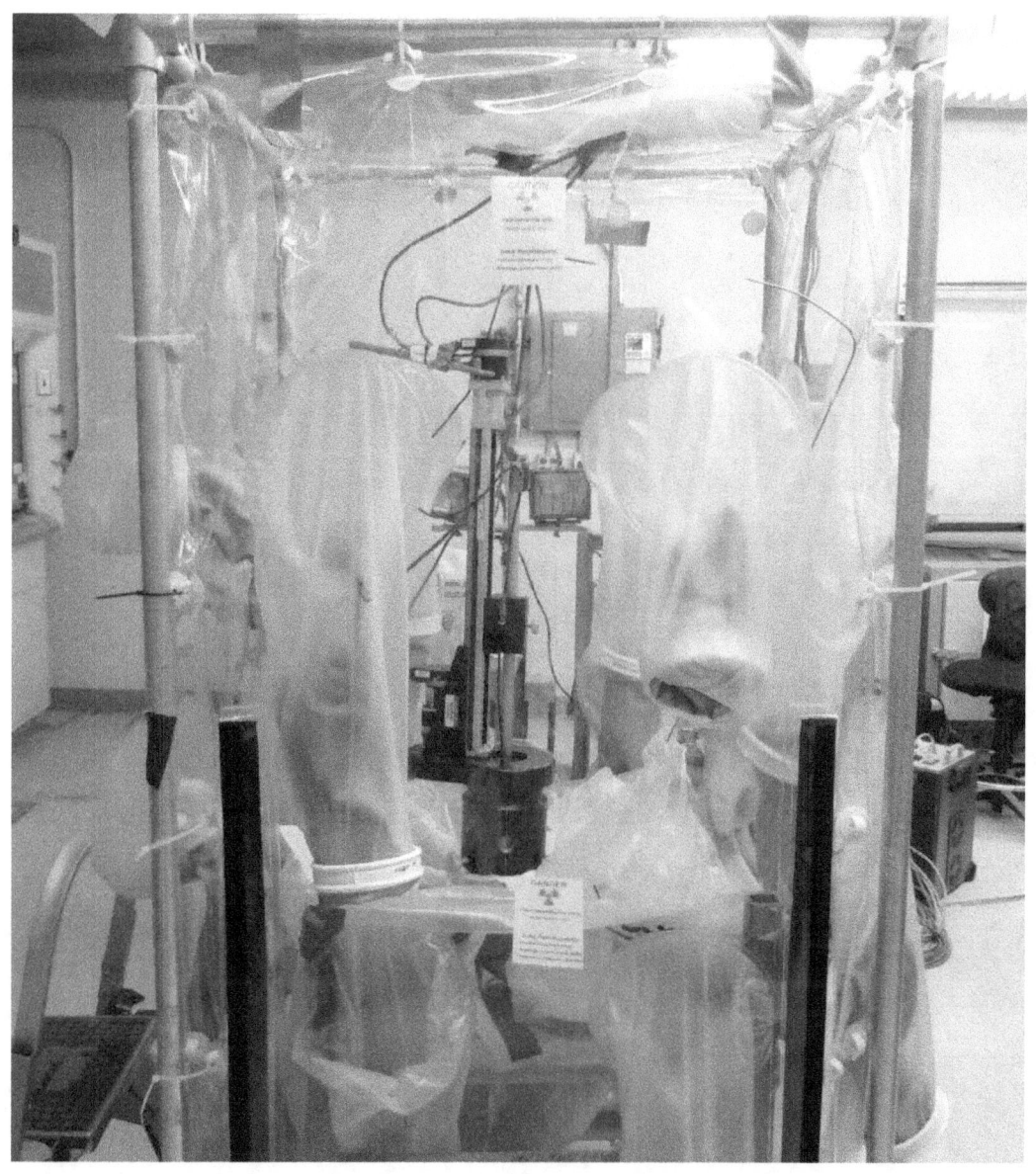

Figure 3.9 Scanner System on Nozzle 63 in the Custom Glovebag

4 CALIBRATION MOCKUP

Prior to performing the non-destructive leak path assessment of Nozzle 63, a control rod drive mechanism (CRDM) mockup was constructed at Pacific Northwest National Laboratory (PNNL) for calibration of the ultrasonic testing equipment and to assess its capability to identify features associated with primary water leakage. These include crystalline boric acid in the interference fit region, wastage or corrosion of the low-alloy steel reactor vessel head material, and cracking or degradation of the Alloy 600 nozzle material. A description of the mockup and testing is presented in this section.

4.1 Mockup Design and Fabrication

The CRDM mockup was made from an Alloy 600 tube fitted with two 152-mm (6-in.) thick low-alloy steel blocks. The mockup was designed to simulate the interference fit between the CRDM nozzle and the reactor pressure vessel (RPV) head material in a pressurized water reactor (PWR), using similar materials and fabrication techniques. The mockup had two interference fit regions as shown in Figure 4.1. In the top interference fit, notches were made on the outer diameter (OD) of the tube and on the low-alloy steel blocks with electric discharge machining (EDM) to simulate cracking, wastage, and degradation of the materials. In the bottom interference fit, crystalline boric acid was placed between the tube and the low-alloy steel blocks to simulate deposits left by primary water leakage. To prevent tipping of the mockup, the flange end of the tube was bolted to a larger plate for increased stability.

4.1.1 Simulated Boric Acid Deposits

The lower interference fit on the CRDM nozzle mockup contained crystalline boric acid deposits in the region between the Alloy 600 tube and the low-alloy steel block. Boric acid deposits in the interference fit of an operating plant could indicate leakage of borated primary water through the J-groove seal weld. In-service inspection data show that the presence of boric acid creates a unique ultrasonic transmission and reflection patterns in the fit regions (Cumblidge et al. 2009). The mockup was designed so that PNNL could evaluate and quantify this ultrasonic transmission and reflection phenomenon prior to examining Nozzle 63.

The lower interference fit mockup region was designed to have both regions where boric acid deposits were present and bare metal regions without boric acid, as shown in Figure 4.2. Ideally, the differences in ultrasonic transmission and reflection for the respective regions would be identified by the test system. The process of creating the boric acid deposit regions began with masking off regions with tape on the Alloy 600 tube OD to preserve the bare metal surface, as shown in the left hand side of Figure 4.3. Simulated boric acid deposits were prepared by mixing a small amount of boric acid in solid form with a small amount of methanol. The two components were then sonicated into a paste with medium to high viscosity. A thin and even coat of the paste was applied with a chemically compatible brush over the localized region on the OD of the tube between the masked-off sections. Upon evaporation of the methanol and solidification of the boric acid, the masking tape was removed. A snake-like pattern was scraped into one of the boric acid regions as indicated with the blue line in Figure 4.2 and the arrow in the right hand side of Figure 4.3.

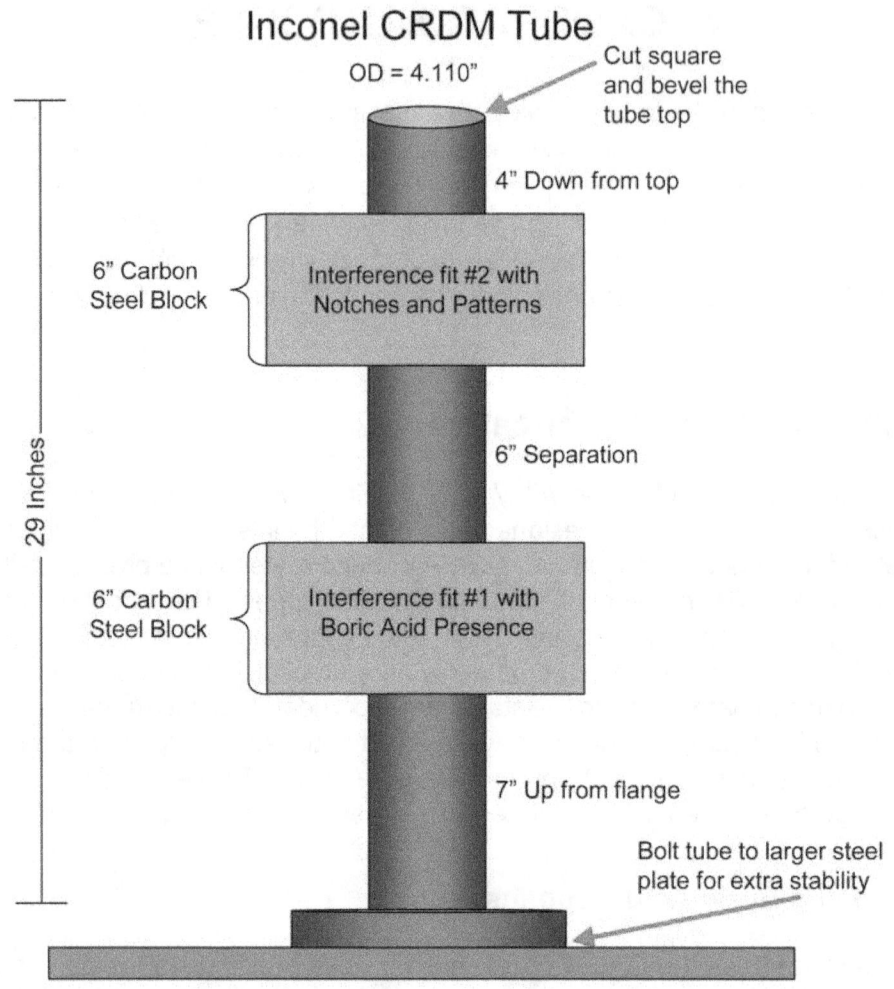

Figure 4.1 Assembled CRDM Interference Fit Mockup Specimen

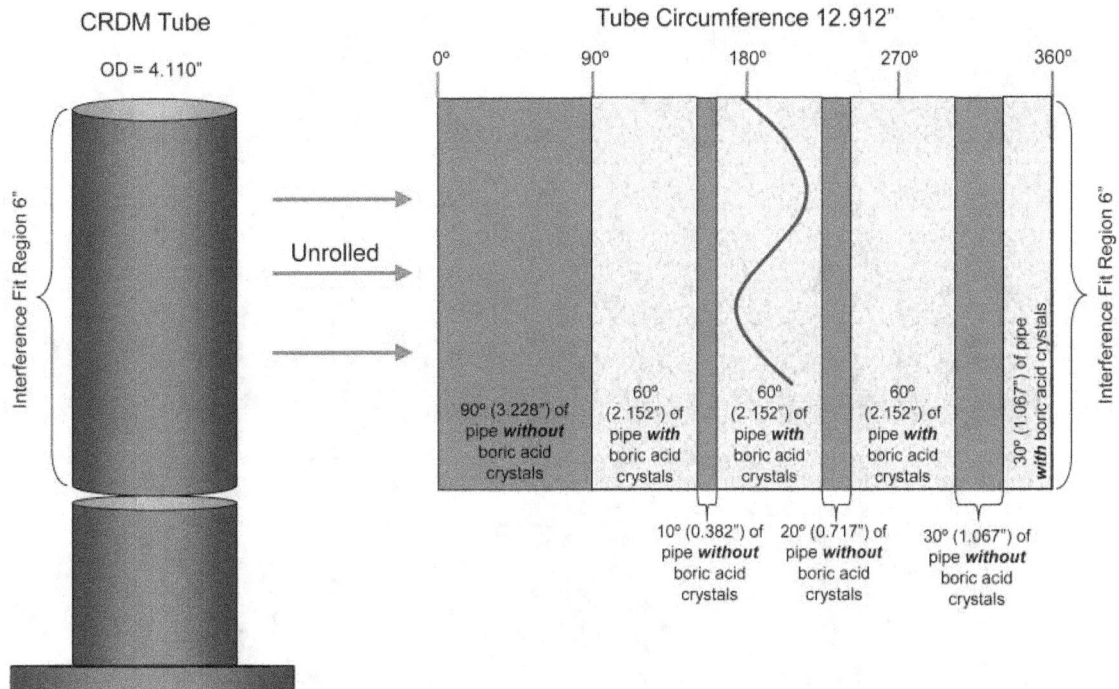

Figure 4.2 Boric Acid Pattern Conceptual Design

**Figure 4.3   Boric Acid Application. Left: Blue painter's tape was used to preserve the
bare metal surface of the Alloy 600 tube. Right: The arrow indicates the path
where boric acid was scraped off.**

4.1.2   Simulated Cracking, Cutting and Wastage

The upper interference fit in the CRDM nozzle mockup contained various precision-crafted EDM
notches to create a small air gap or void region between the tube and the low-alloy steel block.
This was intended to simulate regions in the interference fit where an air gap or path for boric
acid leakage could exist between the penetration and RPV head from wastage of the low-alloy
steel RPV or by anomalies in the CRDM tube such as machining marks, cracking, and steam
cutting. The notches were machined by Western Professional, Inc., with the pattern shown in
Figure 4.4, to provide ultrasonic detection limits and characterization information for voids in the
interference fit region.

As shown in Figure 4.4, notches were put in both the Alloy 600 tube, which is the silver-colored
region in the figure, and the low-alloy steel block, which is the brown/orange colored region.
The tube and the low-alloy steel block had the same notch pattern, but the mockup was oriented
so that the region with notches on the tube did not overlap the region with notches on the block.
The first 180 degrees of the mockup had the notches cut into the tube OD, and the area from

180 to 360 degrees having the notches cut into the low-alloy steel block inner diameter (ID). The notches were oriented both horizontally and vertically to assess probe resolution in the axial and circumferential directions, respectively. A theoretically determined spot size using the 5-MHz phased-array probe at the interference fit region is 1.0 mm (0.04 in.) in both theta and Z directions (circumferential and axial directions). For reference, the theoretical wavelength (λ) in the Alloy 600 tube material at 5 MHz is 1.1 mm (0.043 in.).

The probe resolution in both the circumferential and axial directions was measured by acquiring data on a series of like-sized notches (2 mm wide × 2 mm deep × 25 mm long [0.079 in. × 0.079 in. × 1.0 in.]) that were spaced 2, 3, and 5 mm (0.079, 0.12, and 0.20 in.) apart (approximately 2, 3, and 5 λ), respectively. One set of these notches was orientated circumferentially and the other was oriented axially, as represented by blue lines in the upper left area of the rectangular schematics in Figure 4.4. To measure width detection sensitivity, axially oriented notches with similar length and depth but varying width (0.79–6.35 mm [0.03–0.25 in.]) were machined into the tube and low-alloy steel block, as represented by the lines labeled 1 through 4 in the lower end of the schematics in Figure 4.4. To assess depth sensitivity, a series of axially oriented notches with similar length and width but varying depth (0.025–0.127 mm [0.001–0.005 in.]) were machined into the tube and low-alloy steel block, as represented by the lines labeled 5 through 8 in the upper right of the schematics in Figure 4.4. Complete as-built dimensional details for all of the notches can be found in Appendix A.

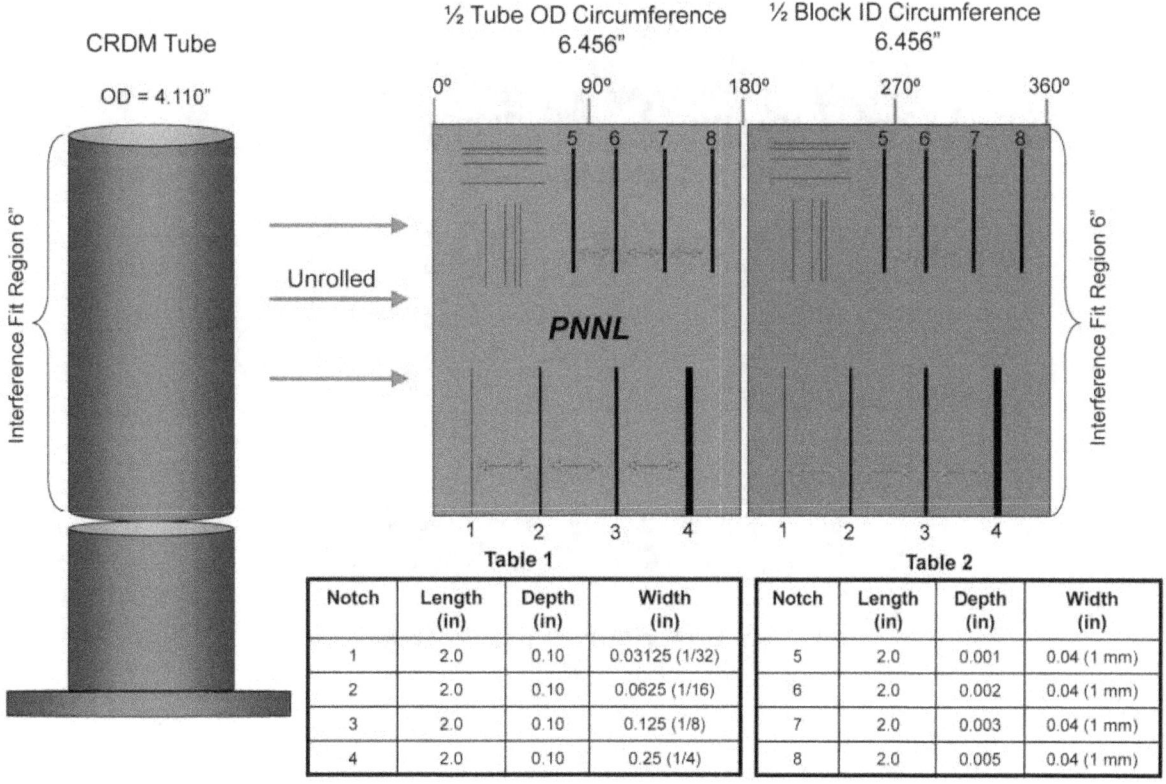

Notch	Length (in)	Depth (in)	Width (in)
1	2.0	0.10	0.03125 (1/32)
2	2.0	0.10	0.0625 (1/16)
3	2.0	0.10	0.125 (1/8)
4	2.0	0.10	0.25 (1/4)

Notch	Length (in)	Depth (in)	Width (in)
5	2.0	0.001	0.04 (1 mm)
6	2.0	0.002	0.04 (1 mm)
7	2.0	0.003	0.04 (1 mm)
8	2.0	0.005	0.04 (1 mm)

Figure 4.4 Interference Fit #2; Notch and Pattern Conceptual Design

The three sets or groupings of notches did not overlap, but were separated so that ultrasonic observations could be made independently on the ability to resolve two closely spaced indications, as well as width and depth sensitivity. The acronym 'PNNL' was also inscribed on the OD of the tube to provide an indication of off-axis sensitivity. Figures 4.5 and 4.6 display the EDM notches in the Alloy 600 tube and low-alloy steel block.

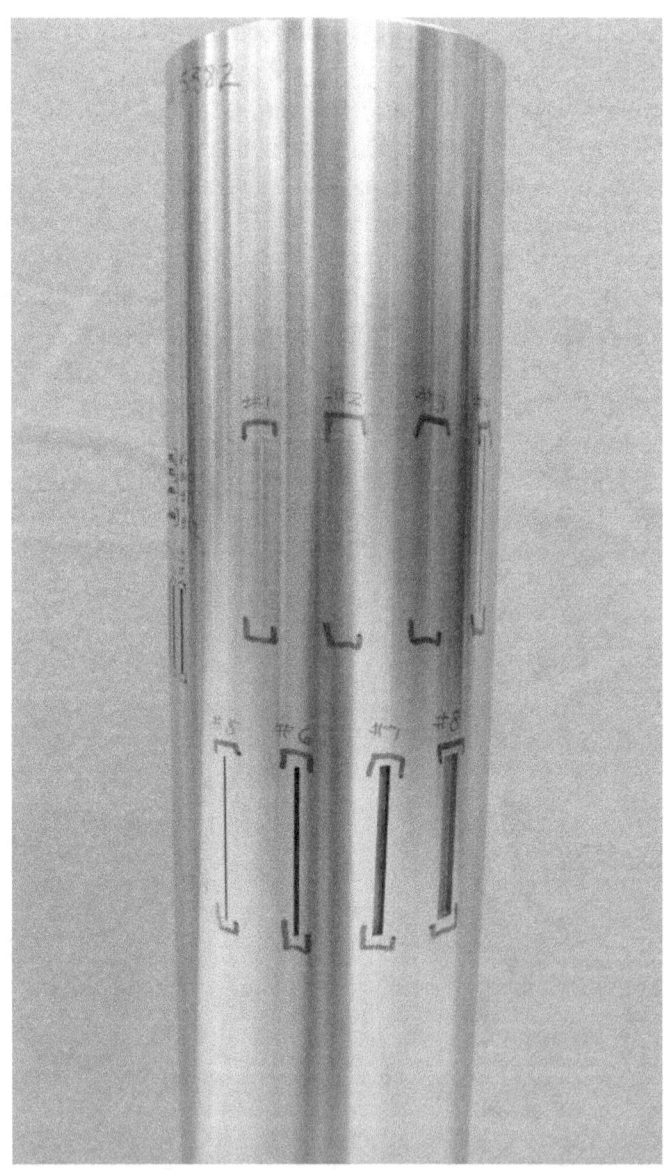

Figure 4.5 EDM Notches in Alloy 600 Tube

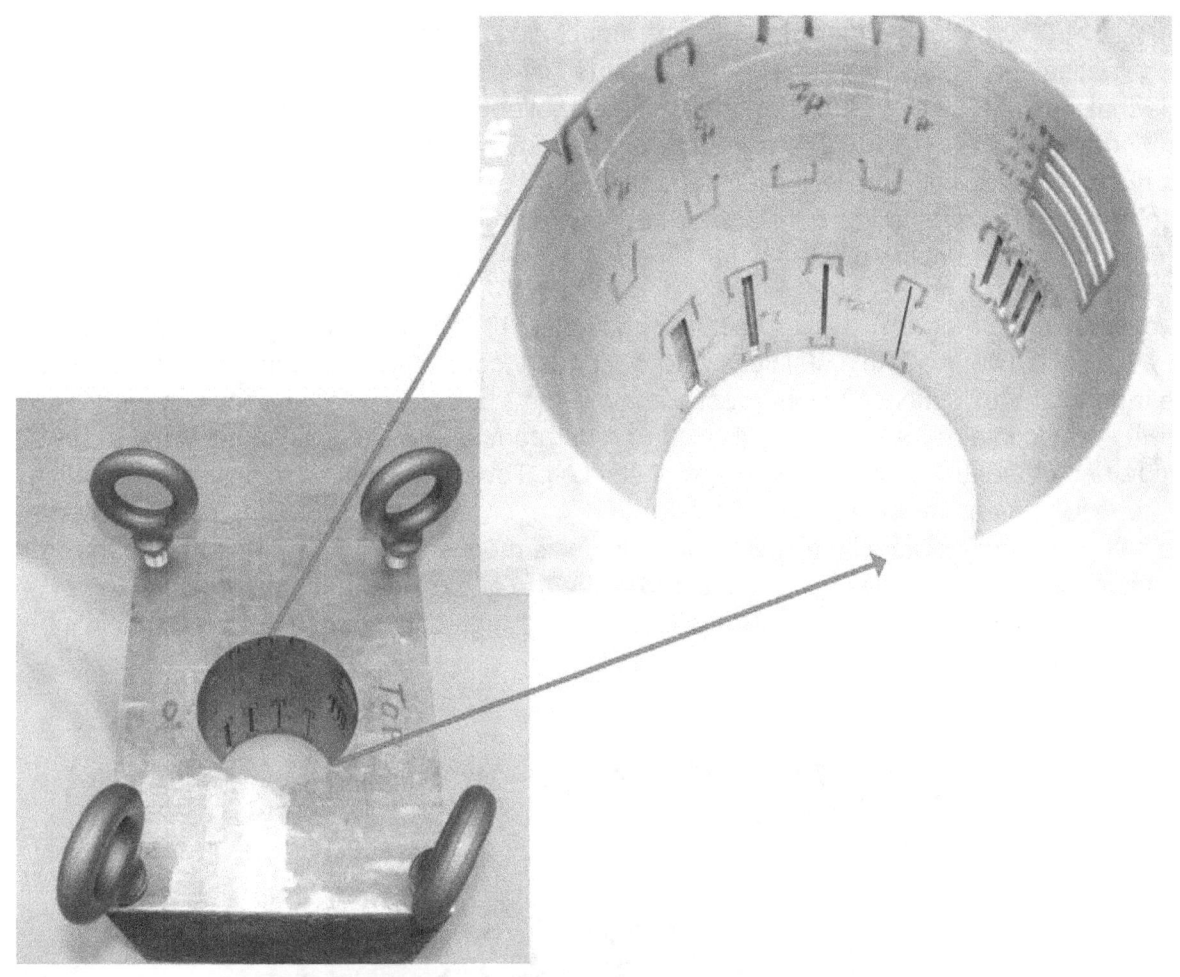

Figure 4.6 EDM Notches on Low-Alloy Steel Block

4.1.3   Mockup Assembly

During RPV head fabrication, nozzle penetrations such as the CRDM nozzle are inserted through holes machined in the RPV head. At room temperature, the diameter of the hole in the RPV head is smaller than the OD of the penetration, creating the compression or interference fit. Therefore, to fit the penetration through the hole in the RPV head, either the RPV head must be heated to increase the diameter of the hole by thermal expansion or the Alloy 600 penetration must be cooled to shrink its diameter, or both. PNNL choose to cool the tube to fabricate the mockup CRDM nozzle. The other parameter considered for fabricating the mockup was the interference fit diameter; this is the amount by which the diameter of the hole is smaller than the OD of the penetration at room temperature. A suggested maximum fit in literature was 0.102 mm (0.004 in.) (Gorman et al. 2009). Reported industry interference fit ranges were listed as 0.030 to 0.102 mm (0.0012 to 0.004 in.) (Hunt and Fleming 2002). PNNL fabricated nominal 0.0762-mm (0.003-in.) interference fits to remain within the industry standard range of 0.030 to 0.102 mm (0.0012 to 0.004 in.).

The Alloy 600 tube for the mockup was lightly machined to remove any minor surface irregularities and its OD was measured. Then the low-alloy steel blocks were machined with a hole that was 0.0762 mm (0.003 in.) in diameter smaller than the OD of the tube at room temperature of 22°C (72°F). The assembly of the CRDM mockup involved temporarily cold-shrinking the Alloy 600 tube with liquid nitrogen (LN), so that it could be fitted with the low-alloy steel blocks. This created an interference fit of 0.0762 mm (0.003 in.) after all components returned to room temperature.

To assemble the mockup, LN was used to shrink the 740-mm (29-in.)-long Alloy 600 tube. As the tube rested vertically in a stainless steel secondary containment trough, LN was added in the tube to approximately 100 mm (approximately 4 in.) below the top, as shown in Figure 4.7. The LN was contained solely within the tube. A permanent end cap was seal welded at the flange end of the tube to prevent leakage of any LN. Towels were used to assist in insulating the tube to prevent unwanted heat transfer and/or ice formation on the OD of the tube. Once the tube cavity was full of LN, the OD of the tube was monitored with calipers until the maximum shrinkage level was achieved. As measured at the top of the tube, a diameter-shrinkage of 0.20 mm (0.0079 in.) was achieved. Additional details regarding shrinkage fits are found in Appendix B.

Figure 4.7 Filling Alloy 600 Tube with Liquid Nitrogen

The first fit to be made was the lower one with simulated boric acid deposits. Oversized polyvinyl chloride (PVC) piping was cut to length and fitted over the Alloy 600 tube and served as a hard stop for the low-alloy steel block to rest on while the specimen returned to room temperature, as seen in Figure 4.8. Next, the insulation towels were removed and the machined low-alloy steel block was hoisted over top of the tube and aligned accordingly. The block was lowered rapidly and slid down the Alloy 600 tube, but came to rest approximately 64 mm (2.5 in.) above the targeted resting place. Thus, the boric acid deposits were only under the bottom half of the low-alloy-steel block. Upon return to room temperature, the PVC piping was no longer needed and was removed.

Figure 4.8 PVC Spacer Shown at Bottom of Specimen

The second interference fit with machined notches was created following a similar protocol. For this fit, it was critical to align the zero degree point of the low-alloy steel block with the zero degree point stamped on the Alloy 600 tube so as to not overlap the notch patterns created in the two materials. The assembly of this fit went according to plan using the PVC piping separator during assembly to maintain separation between the two fit regions. Figure 4.9 shows the completed and assembled mockup.

Figure 4.9 Assembled Mockup

4.2   Ultrasonic Evaluation of Mockup

The CRDM nozzle mockup was examined with the annular ultrasonic phased-array probe described in Section 2. The results of the mockup examination are presented in the following sections.

4.2.1   Alloy 600 Tube Notches

The machined notches in the Alloy 600 tube of the CRDM nozzle mockup simulated air gaps between the penetration and the RPV head that could arise from cracking of the penetration. The notched area shown in Figure 4.4 was scanned over approximately a 0 to 170 degree range in the circumferential direction (horizontal axis) and 0 to 180 mm (7.1 in.) in the axial direction (vertical axis) with the data image shown in Figure 4.10. This top view, plan view, or C-scan image shows the resolution notches in the upper-left portion of the image. The variable depth and width notches are also seen as well as the letters "PNNL." The color scale is displayed on the left with lowest amplitudes at the bottom represented by white and the highest amplitude at the top of the color bar represented by red. In this pulse-echo data, the low amplitude signals (blue and green) indicate good transmission or low reflection of the ultrasonic energy at the interface of the tube to the low-alloy steel. Conversely, the high-amplitude signals (yellow and red) represent poor transmission or large reflection at the interface. A large reflected signal would be generated at a tube-to-air interface as would be seen above and below the interference fit region or in the presence of a notch with large enough dimensions. The system gain level was selected to produce a strong but not saturated response from the largest notch. A full-screen height (FSH) value of approximately 97%, from a possible range of 0 to 100%, is ideal and allows for mapping of the ultrasonic responses over the entire system dynamic range.

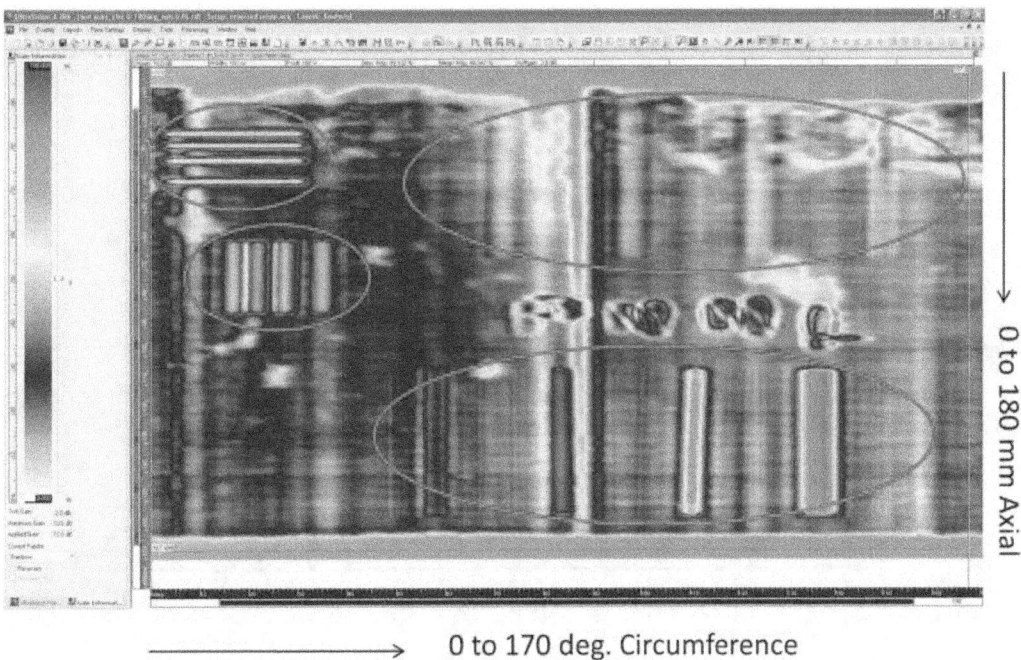

0 to 180 mm Axial

0 to 170 deg. Circumference

Figure 4.10 Top View, Plan View, or C-scan Ultrasonic Image of the Upper Interference Fit Region Containing Calibration Notches in the Alloy 600 Tube. Calibration notches are circled.

The data analysis software allowed electronic gating of signals in the time or depth dimension as well as positional dimensions Z and theta (axial and circumferential, respectively). The axial resolution notches were first gated or selected for analysis. An enlarged D-scan end view, as depicted in Figure 4.11 was used to measure the center-to-center spacing of the notches. This image was taken as viewed from the left edge of the image in Figure 4.10 and depicts depth into the material (along the sound path) in the vertical axis and the scanner index or nozzle axial direction in the horizontal axis. From this end view, an "echodynamic" curve or profile was generated along the red horizontal line drawn through the responses from the notches, and was plotted above the image. The measured notch widths, from left to right, as measured at the half-amplitude points, were 2.0, 2.0, 2.5, and 2.5 mm (0.08, 0.08, 0.10, and 0.10 in.), respectively. This represents only 4 or 5 pixels with each pixel equal to 0.5 mm (0.02 in.). The actual notch widths were 2.08, 2.06, 2.16, and 2.11 mm (0.082, 0.081, 0.085, and 0.083 in.), respectively. Notch depths were measured as 2.06, 2.06, 2.03, and 2.03 mm (0.081, 0.081, 0.080, and 0.080 in.), respectively. Actual depths were 2.06, 1.95, 2.00, and 2.00 mm (0.081,

0.077, 0.079, and 0.079 in.), respectively. The data suggested that gaps created by cracks as small as approximately 1 mm (0.04 in.) in depth could be accurately sized. Also, the notches could be clearly distinguished from each other, providing an indication of lateral probe resolution in the nozzle axial direction. In this set of notches, the actual center-to-center separations were 7.11, 5.08, and 4.06 mm (0.28, 0.20, and 0.16 in.), respectively. The measurements from the ultrasonic data gave separations of 7.0, 5.5, and 4.0 mm (0.28, 0.22, and 0.16 in.), respectively. These highly correlated data values and the data image indicated that an axial resolution of better than 4.0 mm (0.16 in.) was achievable.

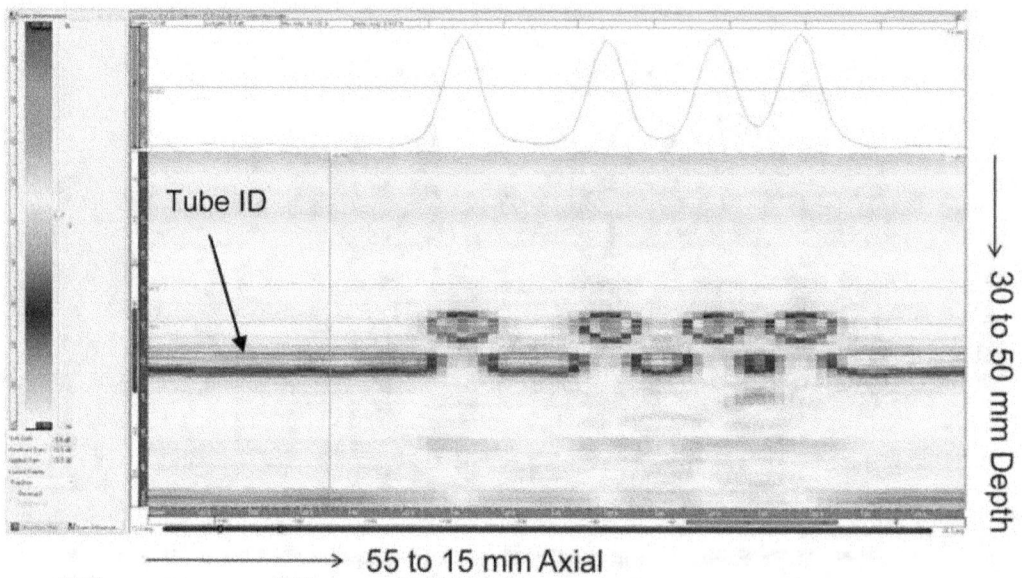

Figure 4.11 D-scan End View of the Axial Resolution Notches in the Alloy 600 Tube

The gated circumferential resolution notch set is shown in Figure 4.12. This image was taken as viewed from the bottom edge in Figure 4.10. Notice that the closely spaced two notches on the left are overlapping but they are still resolvable. Peak-to-peak values gave measured notch separations of 4.36, 5.18, and 6.82 mm (0.17, 0.20, and 0.27 in.), respectively. The actual separations were 4.06, 5.08, and 7.11 mm (0.16, 0.20, and 0.28 in.), respectively. This is greater error than was observed for the axial direction. This test demonstrated a circumferential probe resolution of approximately 4.4 mm (0.17 in.).

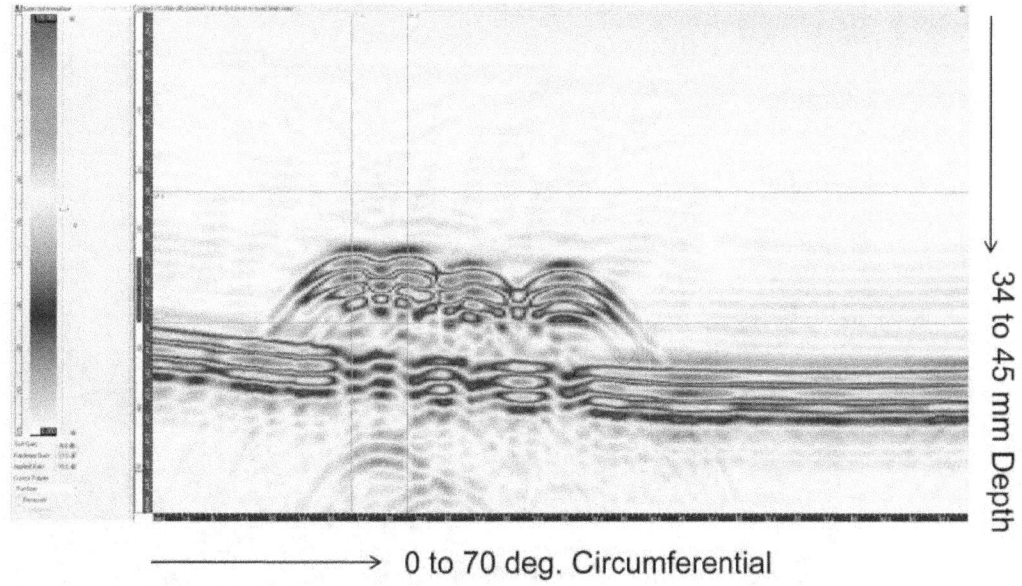

34 to 45 mm Depth

0 to 70 deg. Circumferential

Figure 4.12 B-scan Side View of the Circumferential Resolution Notches in the Alloy 600 Tube

The set of notches in the upper right portion of the scanned image in Figure 4.10 varied in depth but had constant width. These very shallow notches were recognized because their shape and location were known, but they could have been missed based on amplitude response alone. Machining marks on the tube as well as variations in the interference fit produced a non-uniform background response for the fit region, complicating the detection. The center-to-center separations of these notches as ultrasonically measured were 23.84, 24.29, and 23.61 mm (0.939, 0.956, and 0.930 in.), respectively, whereas the actual spacing was 25.4 mm (1.0 in.) between each notch. Notch depth information was not discernible in the first interference fit echo, but the second echo gave some indication of a notch tip as noted by the red arrows in the upper part of Figure 4.13. This image represents the B-scan side view of the data while the lower image is a C-scan top view. A higher inspection frequency could have better resolved the small depth variations in these notches. The current second-echo ultrasonic data showed an approximate depth of 0.15 mm (0.006 in.) for all four notches, whereas the actual depths were 0.028, 0.051, 0.76, and 0.13 mm (0.001, 0.002, 0.003, and 0.005 in.), respectively. While these very shallow notches each presented a discontinuity that was ultrasonically detected, their depths were below the system depth or range resolution. For a greater than 50% bandwidth probe, the range resolution is on the order of one wavelength, which in Alloy 600 is approximately 1.1 mm (0.043 in.) at a 5-MHz inspection frequency.

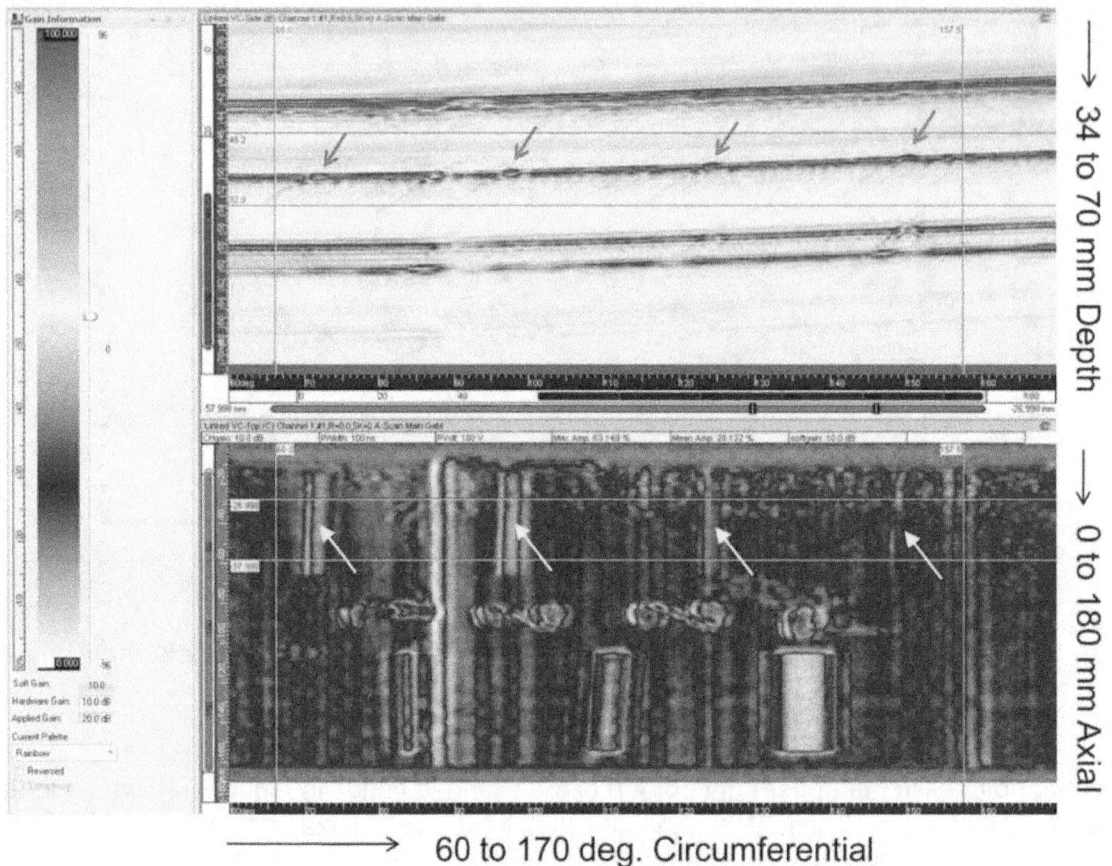

34 to 70 mm Depth

0 to 180 mm Axial

60 to 170 deg. Circumferential

Figure 4.13 The Second Echo is Gated in the Side View Image in the Top of Figure with the Horizontal Lines. The corresponding C-scan top view is displayed in the bottom image. This second echo captures a disturbance in the back-wall echo showing some depth information, noted by the red arrows at top. The yellow arrows note the depth-varying notches.

The final set of notches contained width variations and are shown in Figure 4.14. These notches were ultrasonically measured with depths of 2.2, 2.5, 2.7, and 2.9 mm (0.09, 0.10, 0.11, and 0.11 in.), respectively, left to right in the image, while the actual depth was 2.53 mm (0.10 in.) for all notches. The measured center-to-center spacings were 23.1, 24.8, and 23.8 mm (0.91, 0.98, and 0.94 in.), respectively, while actual spacings were all 24.5 mm (1.00 in.). Finally, the widths of the notches were measured in two ways. The first method used the width of the upper part of the notch response, and the second method used the width of the loss of back-wall signal. The loss of back-wall signal technique was more accurate with measured widths of 3.91, 3.36, 5.00, and 8.82 mm (0.099, 0.126, 0.154, and 0.298 in.), respectively. Actual widths were 0.80, 1.61, 3.24, and 6.42 mm (0.031, 0.063, 0.127, and 0.253 in.), respectively. When measured from the second ultrasonic back-wall echo, the loss of signal measurements gave notch widths of 1.36, 2.73, 3.82, and 7.00 mm (0.054, 0.107, 0.150, and 0.276 in.), which were closer to the actual values. The probe spot size when focused at the interference fit, or 15 mm (0.59 in.) into the Alloy 600, was modeled at 1.2 ×1.2 mm (0.047 ×

0.047 in.) at the −6 dB points. Notch width sizing values are typically oversized by the probe spot size so these measured width values were well within the error expected with this probe.

In summary, the system resolution for detecting air gaps from cracking of the penetration, as represented by notches in the Alloy 600 tube, was better than 4 mm (0.16 in.) in the axial direction and 4.4 mm (0.17 in.) or greater in the circumferential direction. The depth or range resolution notches, as small as 0.028 mm (0.0011 in.), were beyond the system limits for depth sizing but the notches were detected. Range resolution was estimated at 1 mm (0.039 in.). Notches as narrow as 0.80 mm (0.031 in.) in the circumferential direction were detected and sized but the limits were somewhat dependent on the machining marks and other anomalies in the materials and interference fit that also gave ultrasonic indications.

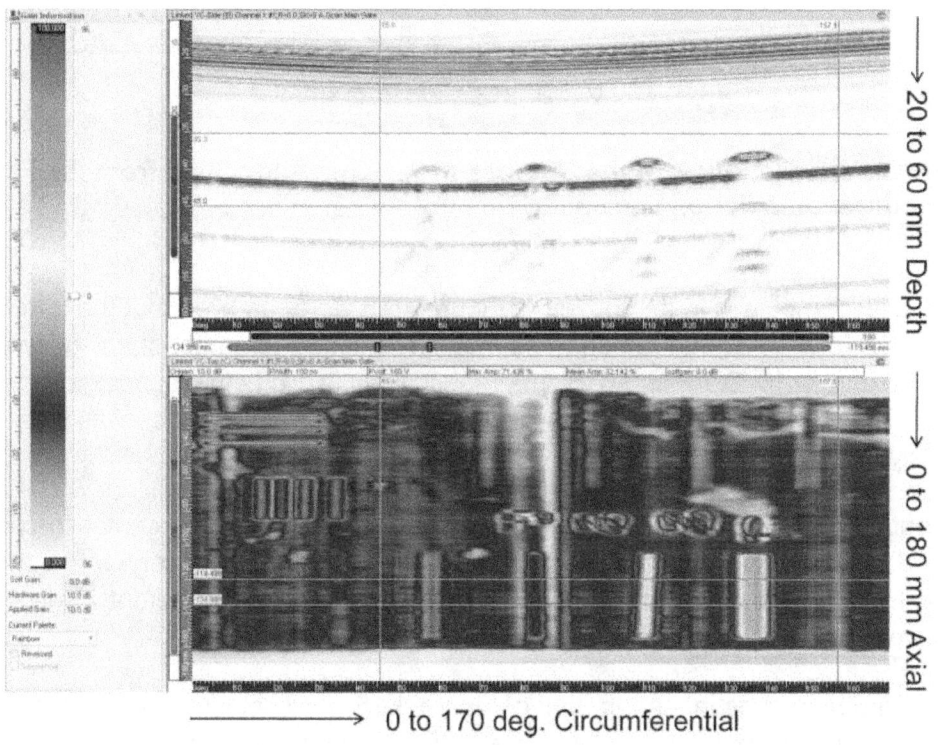

0 to 170 deg. Circumferential

Figure 4.14 B-scan Side View on Top and C-scan Plan View on Bottom of the Width Varying Notches in the Inconel Tube

4.2.2   Low-Alloy Steel Notches

The 180–360 degree portion of the upper fit region in the CRDM mockup contained notches in the low-alloy steel block to simulate an air gap between the penetration and the RPV head created by degradation or wastage of the RPV head material. These notches were on the far side of the interference fit relative to the location of the probe. Because the interference fit was not uniform, the notch responses were not as clear as those for notches in the tube, as evident in comparing Figure 4.15 to Figure 4.10.

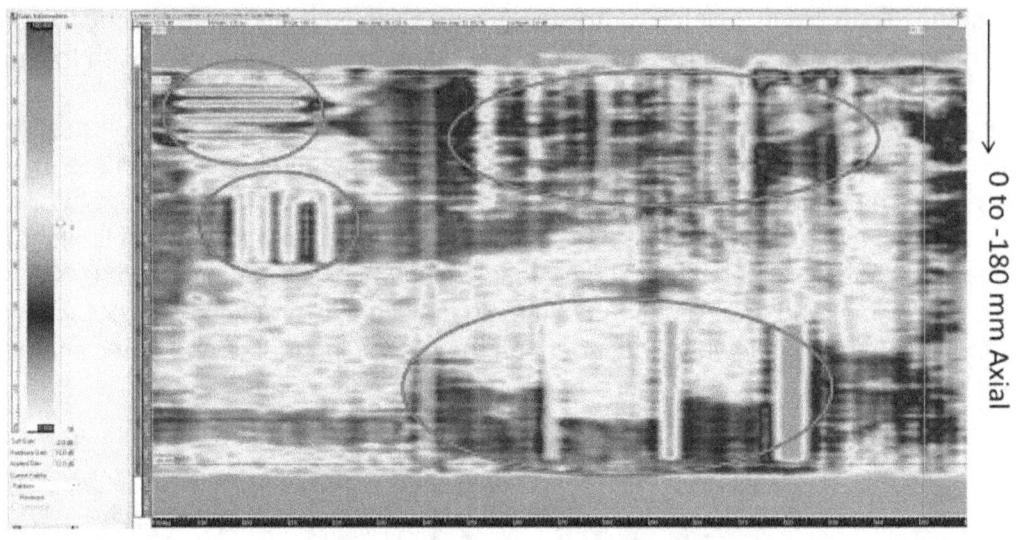

0 to -180 mm Axial

180 to 360 deg. Circumferential

Figure 4.15 C-scan Plan View of the Notches in the Low-Alloy Steel from the First Ultrasonic Echo. The calibration notches are circled.

The axial resolution notches in the top left of Figure 4.15 were resolved, but the lower notch was on the edge of a high-amplitude region. The measured center-to-center spacings were 3.90, 4.58, and 6.88 mm (0.15, 0.18, and 0.27 in.), respectively, while actual spacings were 4.06, 5.08, and 7.11 mm (0.16, 0.20, and 0.28 in.), respectively. Axial resolution was therefore approximately 4 mm (0.16 in.) or better.

Measurements from the circumferential resolution notch pattern showed center-to-center spacings of 3.82, 5.00, and 7.54 mm (0.15, 0.20, and 0.30 in.), respectively, with actual spacings of 4.06, 5.08, and 7.11 mm (0.16, 0.20, and 0.28 in.). Circumferential resolution was also approximately 4 mm (0.16 in.) or better.

The variable depth notches in the top right of Figure 4.15 were detected but depths could not be measured. First and second echo images are shown in Figures 4.16 and 4.17, respectively. Center-to-center spacing was ultrasonically measured at 23.9, 22.4, and 24.4 mm (0.94, 0.88, and 0.96 in.), respectively, with an actual spacing of 25.4 mm (1.00 in.) for all notches.

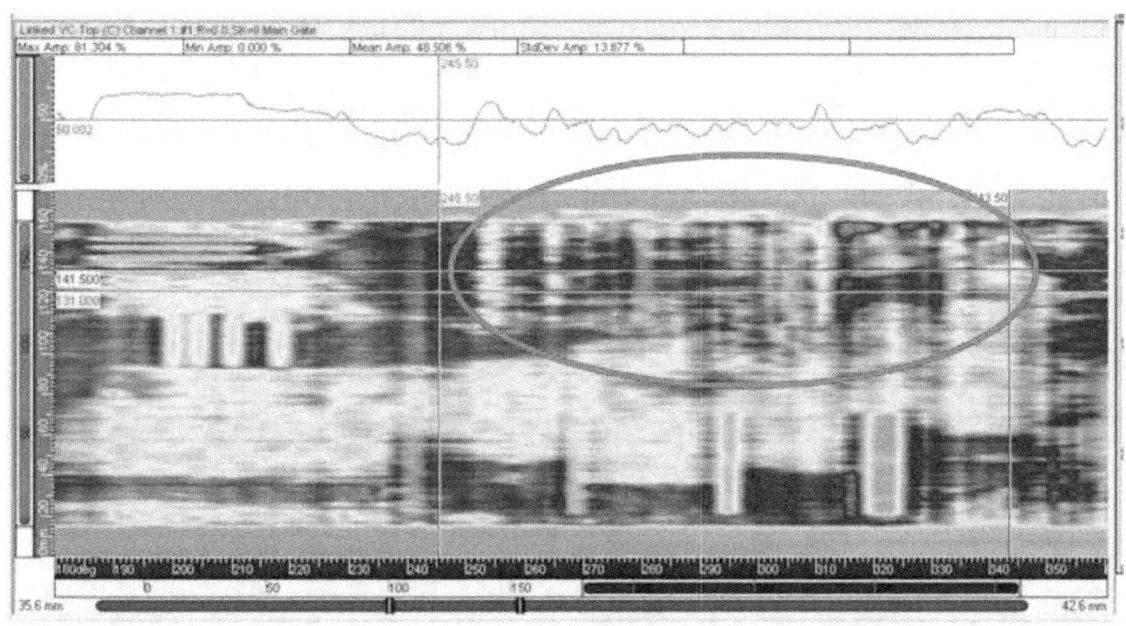

Figure 4.16 C-scan Plan View of the Depth Notches in Low-Alloy Steel, Circled on the Upper Right. This image was acquired from the first ultrasonic echo.

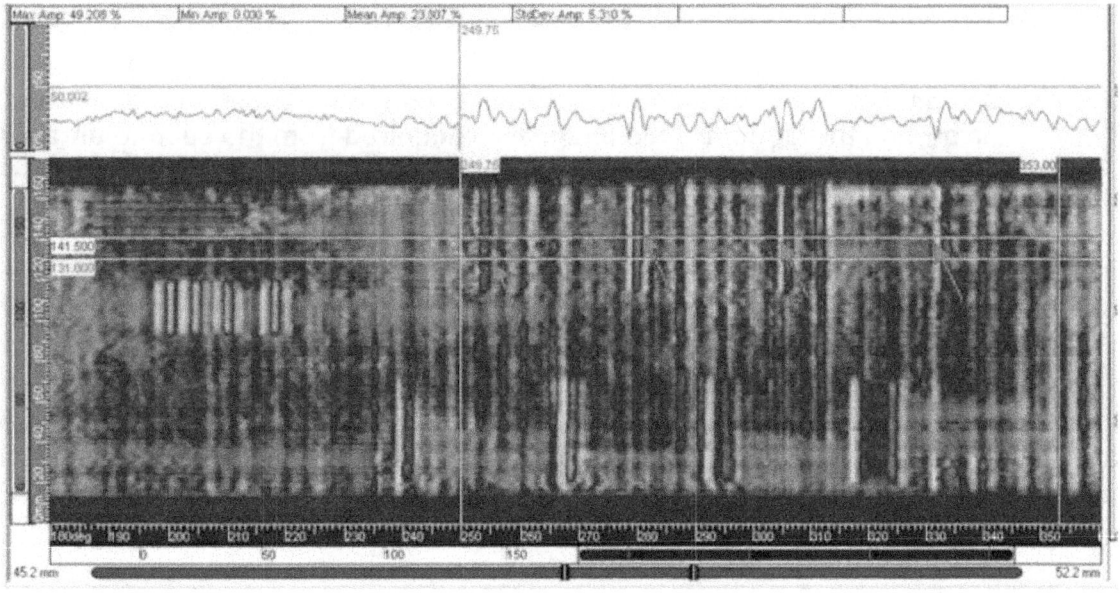

Figure 4.17 C-scan Plan View of the Depth Notches in Low-Alloy Steel, Noted by Red Arrows on the Upper Right. This image was acquired from the second ultrasonic echo.

The set of notches with variable widths is shown in Figure 4.18 with the notches marked by red arrows at the bottom of the image. This image represents the second echo. The center-to-center spacing measurements were 25.4, 23.2, and 25.0 mm (1.00, 0.91, and 0.98 in.), respectively, left to right, with actual spacing of 25.4 mm (1.00 in.). Measured notch width values were 1.82, 2.27, 4.27, and 6.82 mm (0.072, 0.089, 0.168, and 0.268 in.), respectively, with actual values of 0.80, 1.59, 3.18, and 6.39 mm (0.032, 0.062, 0.13, and 0.25 in.), respectively. The notch widths were also measured from the first echo (refer to Figure 4.15, a first ultrasonic echo image) with slightly poorer results.

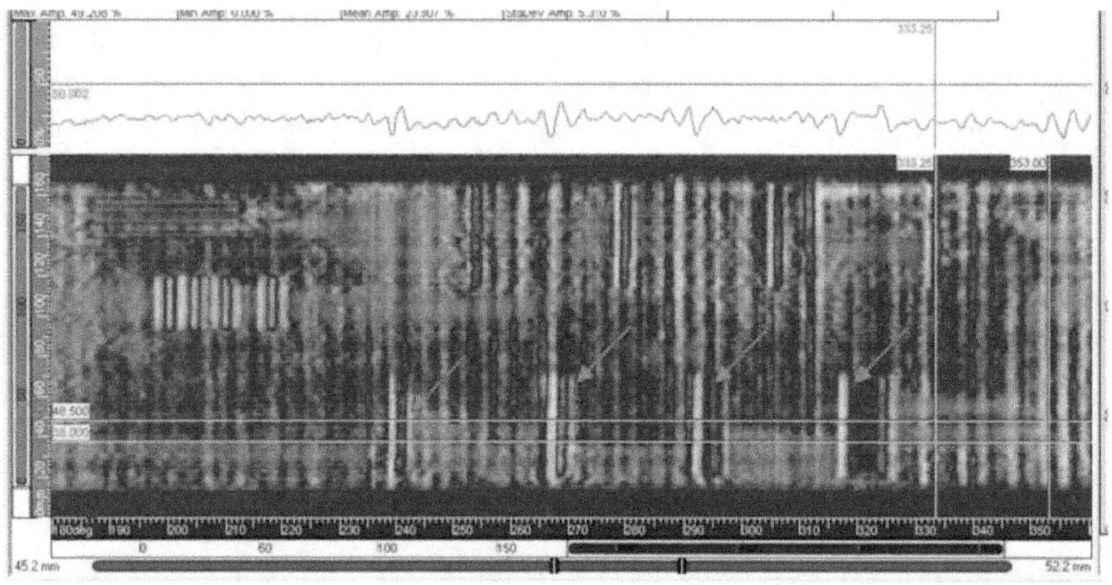

Figure 4.18 C-scan Plan View of the Width Notches in Low-Alloy Steel, Noted by Red Arrows on the Bottom. This image was acquired from the second ultrasonic echo.

In summary, the system resolution for detecting air gaps from corrosion or wastage of the RPV head, as represented by notches in the low-alloy steel block of the mockup was better than 4 mm (0.16 in.) in both the axial and circumferential directions. The depth or range resolution notches, as small as 0.028 mm (0.0011 in.), were beyond the system limits for sizing but the notches were detected. In general, the notch depth into the low-alloy steel is not measureable because the sound beam is reflected at the first tube-to-air interface and does not travel through the air gap to the back of the cavity in the steel. Notches as narrow as 0.80 mm (0.10 in.) in the circumferential direction were detected.

4.2.3 Simulated Boric Acid Deposits

The top view, C-scan images from the mockup with boric acid deposits in the interference fit region between the tube and low-alloy steel block are displayed in Figures 4.19 and 4.20. The first image represents the 60 to 240 degree circumferential region and the second image represents the 240 to 60 degree circumferential region, both as captured by the first echo. The boric acid regions were readily detected as lower amplitude response relative to regions with no boric acid and are outlined with red boxes. Again, notice machining marks and non-uniformity in the interference fit response. The snake path scratched into a boric acid region as visually shown in Figure 4.3 was not evident in the ultrasonic data. Likely, the pattern was not preserved during the mockup assembly.

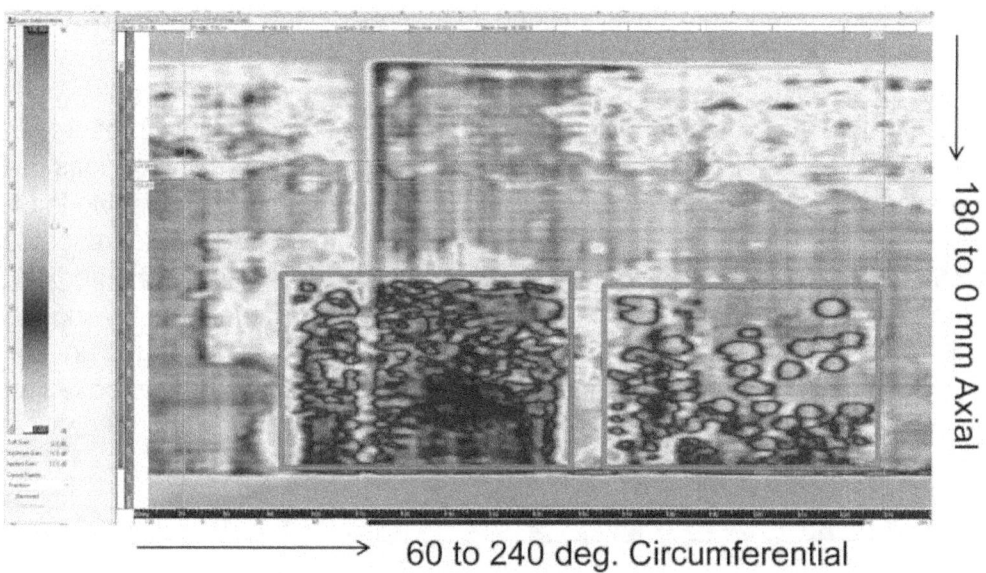

60 to 240 deg. Circumferential

Figure 4.19 C-scan Plan View of the Boric Acid Deposits, Boxed in Red, in the Lower Interference Fit Region. The horizontal axis represents the circumferential range of 60–240 degrees. This image is from the first ultrasonic echo.

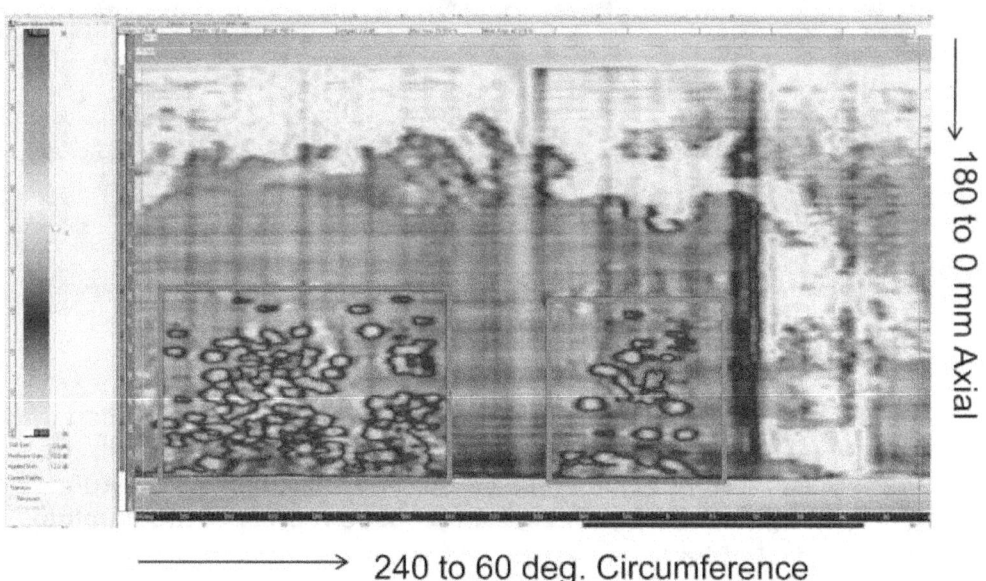

240 to 60 deg. Circumference

Figure 4.20 C-scan Plan View of the Boric Acid Deposits, Boxed in Red, in the Lower Interference Fit Region. The horizontal axis represents the circumferential range of 240–60 degrees. This image is from the first ultrasonic echo.

4.2.4 Amplitude Response

In addition to characterizing the notch response data for probe spatial and range resolution and notch detection capability, an analysis of the acoustic response from the different regions was performed based on the reflected signal strength in the lower boric acid mockup section. Three conditions of the interference fit region are represented in the mockup: the nominal interference fit between the tube and low-alloy steel with no boric acid in between, which are green and yellow areas in Figure 4.21; the interference fit region where boric acid was present between the tube and low-alloy steel block, which are blue areas in Figure 4.21; and the regions of a tube-to-air interface above or below the interference fit, which are orange areas spanning the top and bottom in Figure 4.21. The portion of Figure 4.21 outlined with the red box represents the interference fit region where boric acid deposits were present, the black dashed boxes represent the tube-to-air interface regions above and below the interference fit, and the black dotted boxes represent regions of the nominal interference fit between the tube and low-alloy steel block. The mean and peak amplitudes were measured in each of these boxed areas from the C-scan image. Similar measurements were also acquired for the 240 to 60 degree boric acid image.

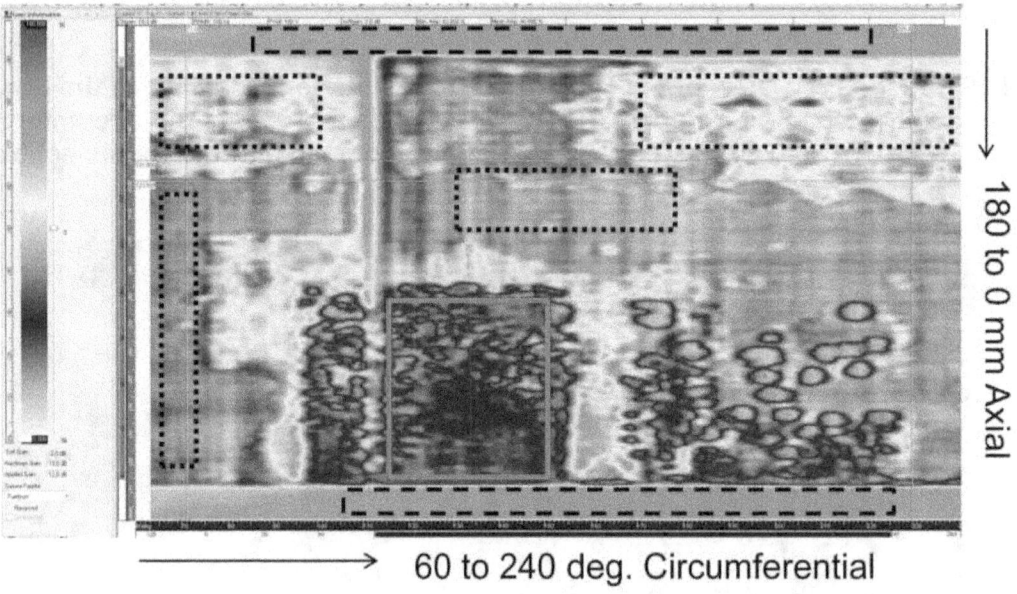

60 to 240 deg. Circumferential

180 to 0 mm Axial

Figure 4.21 The Interference Fit Region Containing Boric Acid is Subdivided into Three Regions. The red box represents the presence of boric acid in the interference fit region, the black dashed boxes represent the tube-to-air interface region, and the black dotted boxes represent the nominal interference fit region.

The data images were analyzed with a total image gain of 12 dB. This represented 10 dB of hard gain applied during acquisition and 2 dB of soft gain applied during analysis. The mean responses from the nominal interference fit regions without boric acid deposits between the tube and low-alloy steel block were in the range of 40 to 55 percent of FSH. This range of values was due to the variability in the fit itself, likely caused by out-of-roundness of the tube or hole in the block. Some regions of the fit were tight, giving lower reflected amplitude and more transmitted energy. This condition is represented by the green color in Figure 4.21. Other regions of the nominal fit with no boric acid showed higher reflected energy indicating less intimate contact, thus less transmitted energy, and this state is represented by the yellow-to-orange colors in Figure 4.21. Machining marks were evident and also lead to response variability. Additionally, the Alloy 600 tubes are known to have hard spots, which could transmit or scatter ultrasonic energy differently than the nominal material. The mean responses of the interference fit regions with boric acid deposits between the tube and low-alloy steel block were in the range of 24 to 30 percent range of FSH. This shows more energy transmitted (less reflected) through the interference fit region with the presence of boric acid than in the regions without boric acid. The boric acid crystals filled gaps in the fit and efficiently coupled the ultrasonic energy from the tube into the low-alloy steel material. Finally, the mean responses of the tube-to-air regions above or below the interference fit were 60 to 75 percent of FSH, demonstrating greater reflectance of energy compared to the interference fit. In summary, these data from the mockup showed that the ultrasonic system could distinguish the condition of the interference fit, whether the nominal state, with boric acid deposits, or an air gap, by their mean ultrasonic response.

A final study was conducted as a result of discussions with John P. Lareau of WesDyne International on industry-style CRDM inspections and practices. He reported that the presence of boric acid was simulated with clay on a nozzle mockup specimen and gives an ultrasonic response that is 2 dB lower than the nozzle without clay. To evaluate the PNNL inspection system under a similar scenario, putty was placed on the outside of a blank nozzle specimen. The presence of putty was clearly seen with the mean response from the putty region measured at 64.8 percent of FSH and the clean nozzle response measured at 72.0 percent. This represents a 0.9 dB drop in amplitude. This smaller response difference is possibly due to the type of clay used in the WesDyne testing as compared to the putty used at PNNL.

5   NOZZLE 63 NONDESTRUCTIVE LEAK PATH ASSESSMENT

After calibration and testing on the control rod drive mechanism (CRDM) mockup, the phased-array ultrasonic equipment was transported to the Radiochemical Processing Laboratory (RPL/33) for the examination of Nozzle 63. The results of the examination are presented in this section.

5.1   Scanner Setup

An Alloy 600 nozzle tube on hand at Pacific Northwest National Laboratory from the canceled Washington Nuclear Power (WNP-1) plant was used to assess the functionality of the probe, scanner, and electronics after they were moved to RPL/33 where Nozzle 63 was housed. A simple scan on the nozzle was performed in the ultrasonic laboratory where the mockup was tested and then in RPL/33 after transporting and reassembling all of the equipment. The maximum and mean amplitude responses were within 0.23 and 0.08 dB, respectively, of each other. Typically calibration data that fall within ±2 dB of each other are acceptable so the equipment functionality was validated.

After equipment verification, the scanner was placed on Nozzle 63 in the glovebag. What was the wetted side of the nozzle in the reactor was facing down in the glovebag. A plug was inserted in the wetted side of the nozzle several months earlier. Water was added several days prior to mounting the scanner on the nozzle. With the scanner in place, the glovebag was examined for any leakage points in the bag walls and glove ports. No leaks were found and the glovebag was sealed. Any materials entering the glovebag from this point forward were passed through an access port in the bag wall.

5.2   Data Acquisition

Before the start of a scan, the probe was manually lowered to the bottom of the interference fit region and the data were collected while indexing the probe upwards. Two data channels were recorded during acquisition—one with a focus at the front surface (1 mm or 0.04 in. within the Alloy 600 material) and the other focused at the interference fit interface (15 mm or 0.59 in. within the Alloy 600 material). The front surface reflection represented the interface between the water inside the tube and the Alloy 600 tube inner diameter (ID), and provided an indication of the surface condition of the tube. This data would reveal bubbles, pitting, and other surface anomalies, if present. The second data channel recorded the reflection from the interface between outer diameter (OD) of the Alloy 600 tube and the low-alloy steel RPV head material, and represented the interference fit region. The first coarse scan data showed the need to laterally adjust the scanner to more accurately center the phased-array probe in the nozzle. Results from data acquired towards the top of the tube before and after centering the scanner are displayed in Figure 5.1. The horizontal axes in each image represent a circumferential distance from −90 to +90 degrees. The vertical axes represent approximately 20 mm (0.79 in.) of travel in depth or distance from the probe. Before centering the scanner, the front surface signal travel or difference from high to low point was measured at 10 mm (0.39 in.) on the left

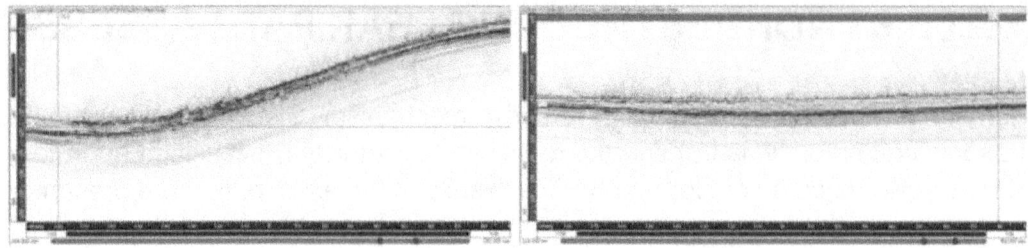

Figure 5.1   Alloy 600 Tube ID Response Before (left) and After (right) Centering the Scanner on the Nozzle. The horizontal axis represents approximately 180 degree and the vertical axis 20 mm (0.79 in.).

image. After centering, the signal travel was only 1 mm (0.04 in.) with the data displayed on the right of Figure 5.1. In addition to showing that the probe was centered, the coarse scans also verified that the areas of interest were captured in the data file.

Once the areas of interest were bounded, the scanner step sizes were reduced for more detailed imaging. A resolution of 0.5 degree in the circumferential (scan) direction and 0.5 mm (0.02 in.) in the axial (index) direction were selected. The ZETEC UltraVision software limits the data files to 1 gigabyte in size. Working within this constraint, data in a file were collected over a range of approximately 180 degrees circumferentially and 380 mm (14.96 in.) axially. As discussed in Section 5.4, an in-service ultrasonic examination of Nozzle 63 by the licensee prior to the head replacement indicated a leakage path at the low point or downhill side of the nozzle, which is the 180-degree location on the coordinate system established for this investigation. Therefore, data were acquired over an approximate −90 to +90 degree region and a 90 to 270 degree region to capture the possible leakage path in the center of an image. The actual circumferential scan regions were slightly larger than 180 degrees to provide some overlap in the data.

The first data covering the 90 to 270 degree area are shown in Figure 5.2. The front surface echo is displayed on the top and the interference fit echo on the bottom. These C-scan top view images show approximately 180 degrees across the horizontal axis and 360 mm (14.17 in.) on the vertical axis. The color bar on the left shows low-amplitude signals in blue/white, which represent good transmission or poor reflectance. High-amplitude signals in orange/red conversely represent poor transmission and good reflectance. The weld region is shown in the white-to-light-blue color at the bottom of the interference fit image. The interference fit or shrink-fit zone is located between the counter bore regions as was shown in Figure 1.1. The data above the interference fit (dry-side annulus region) represent a tube-to-air interface and should provide a strong and uniform reflection. Such a strong reflection was only evident in the orange-colored regions in the right side of the images. The tube OD-to-air interface is a good reflector, so a uniform orange color would be expected across the top of the image. Therefore, the lack of uniformity across the upper portion of the image (tube OD) was unexpected. The lack of uniformity across the entire front surface echo (upper image in the figure) was also unexpected.

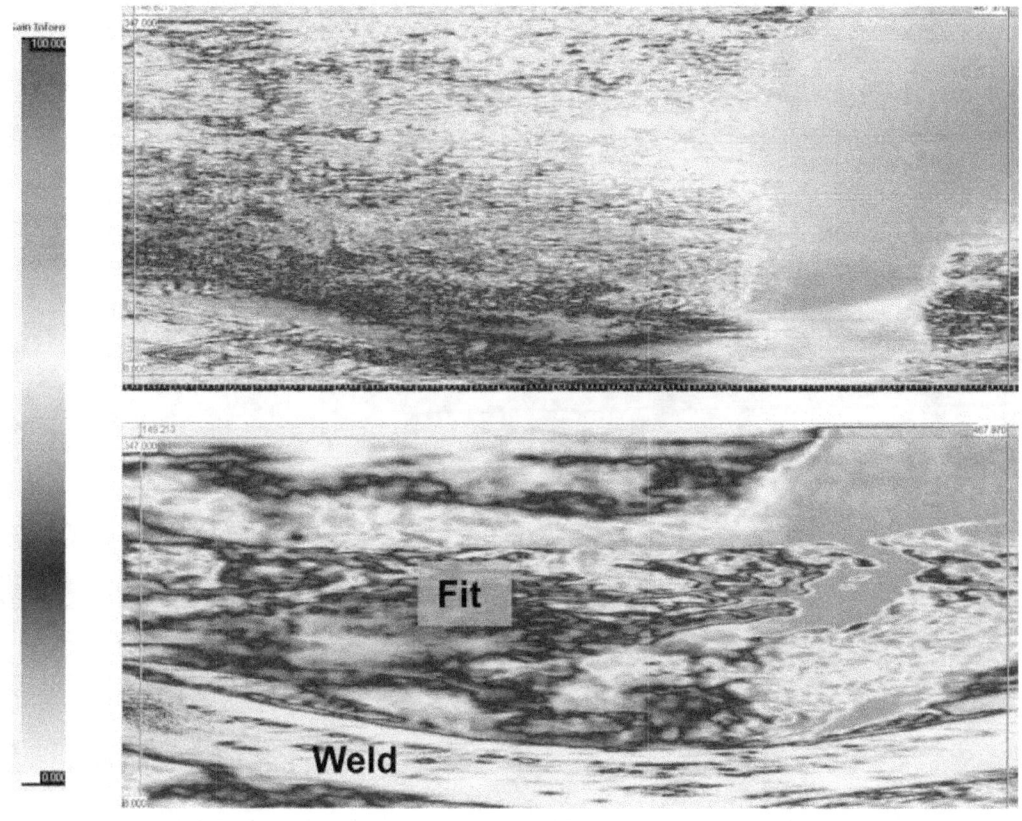

Figure 5.2 First PA Ultrasonic Data from Nozzle 63. The front surface or nozzle ID echo is on the top and the interference fit echo on the bottom. The horizontal axis represents the 86 to 274 degree area and the vertical axis represents 360 mm (14.17 in.). The color scale is represented on the far left.

After the first scanning attempt, the probe was lifted above the water line and found to be dirty. The probe face was carefully wiped with a dry cloth. It was also suspected that bubbles on the ID tube surface could be partly responsible for the degraded image. After carefully brushing the inside of the tube several times, a good data set was obtained.

Data acquired after brushing from the two scans were pieced together to form the composite image of the interference fit region in Figure 5.3. This image displays a full 360-degree representation of the weld and interference fit region with −90 degrees at the left and 270 degrees at the right. The data acquired from the mockup indicated that the nominal interference fit between the penetration and RPV head would appear as green to yellow, boric acid deposits would appear as blue, and a tube-to-air interface or air gap between the penetration and RPV head would appear orange. A suspected gap or leakage path in Nozzle 63 at the low point, near 180 degrees, is marked with arrows in Figure 5.3.

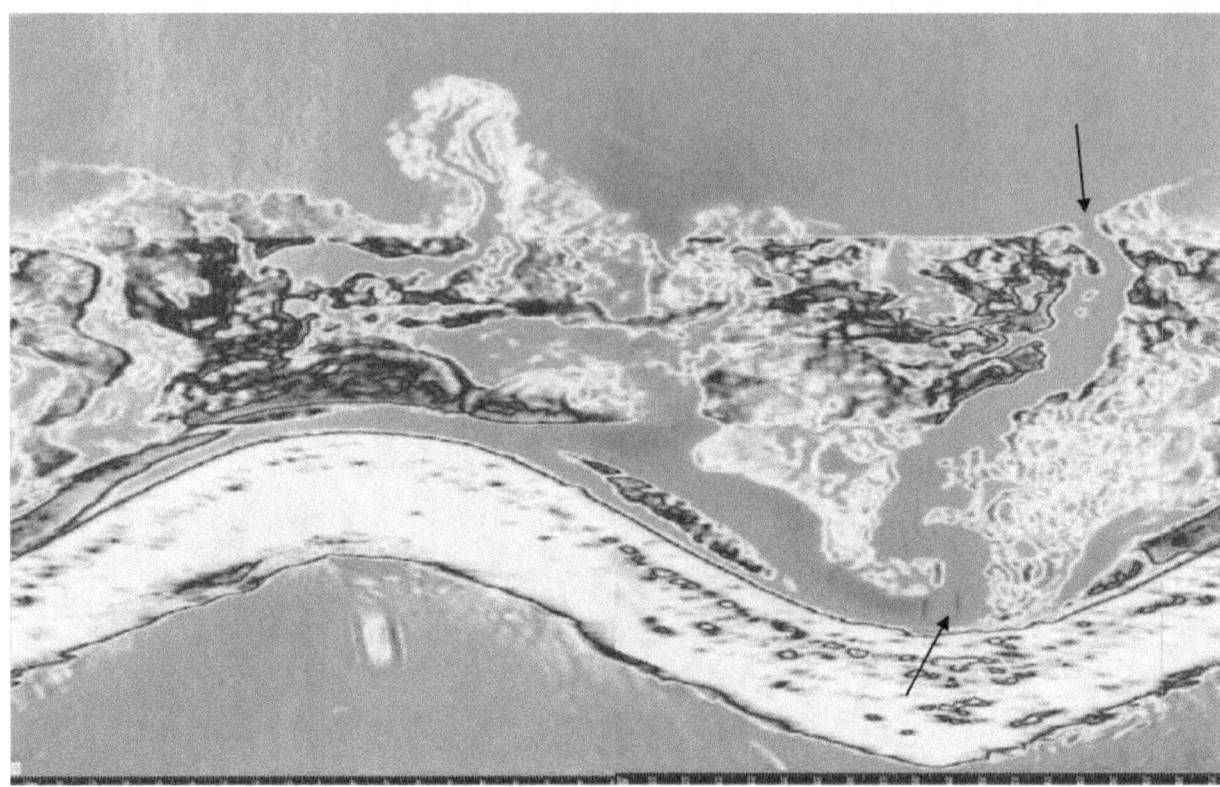

Figure 5.3   PA Ultrasonic Data from the Interference Fit in Nozzle 63 Acquired After the Second Brushing of the Nozzle ID. The horizontal axis represents the full 360-degree area and the vertical axis represents 360 mm (14.17 in.). Black arrows indicate primary leak path. The left edge marks the −90 degree location and the right edge the 270 degree location.

5.3   Amplitude Analysis

Based on the results from the mockup, the image in Figure 5.3 was interpreted to suggest that regions of nominal interference fit between the nozzle penetration and the RPV head would appear green or yellow, that boric acid deposits would appear as blue, and leakage paths and regions above or below the interference fit would appear orange. An analysis was performed to compare the amplitude responses for the Nozzle 63 interference fit to those of the mockup. The first such analysis was conducted on the 90 to 270 degree data acquired after the first brushing. The image was segregated into regions as depicted in Figure 5.4. The peak and mean amplitude responses were measured in each boxed region. Regions 1 through 6 are suspected air gaps or leakage paths in the interference fit between the penetration and RPV head. Regions 7 through 11 represent the response from the tube-to-air interface above the interference fit. The nominal interference fit and lower counter bore areas are represented by regions 19 through 22 while regions 24 through 26 represent the interference fit with suspected boric acid present between the penetration and the RPV head. The same procedure was used on the −90 to +90 degree data with the numbered boxed regions shown in Figure 5.5. Next, the mean amplitudes within each boxed region were plotted with results in Figure 5.6. The suspected leakage path and tube-to-air responses all were in the range of 60 percent of FSH and greater. The regions with suspected boric acid deposits had mean amplitudes 25 percent and below. Finally, the interference fit/counter bore region mean amplitudes were in the 41 to 58 percent range. A proposed segregation of the regions is denoted by the orange, green, and blue colored zones in the plot. A 30 percent mean amplitude or less was thought to indicate the presence of boric acid in the interference fit while greater than 60 percent was thought to represent an air gap and possible leakage path if connected all the way through (top to bottom) the interference fit.

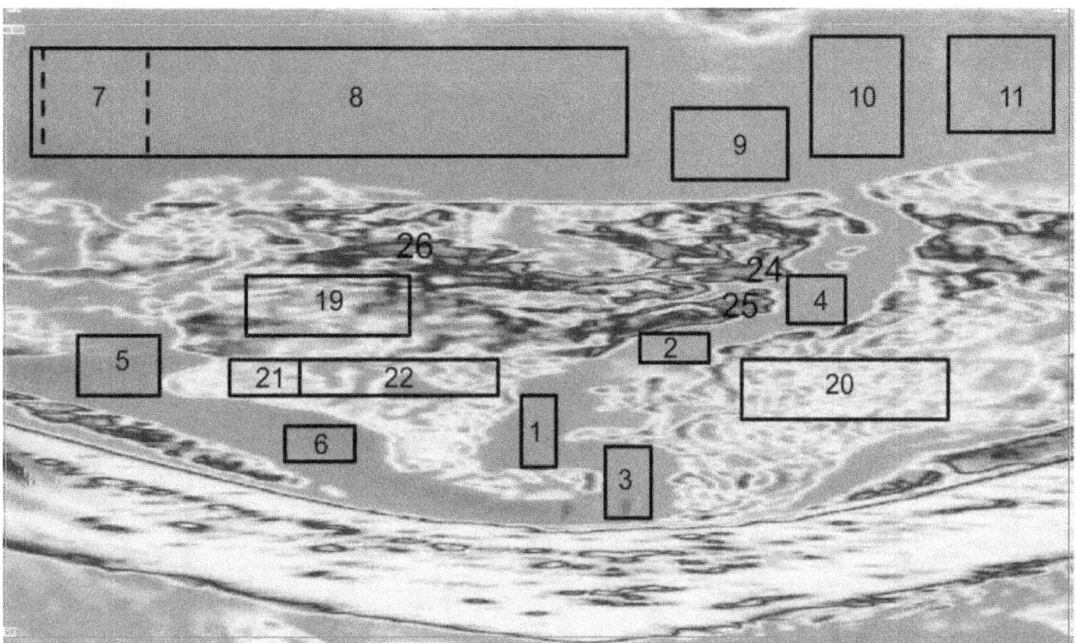

Figure 5.4   Interference Fit Data Image After First Brushing. The horizontal axis represents approximately 90 to 270 degrees. The vertical axis represents 360 mm (14.17 in.). Boxed regions were used in an amplitude analysis.

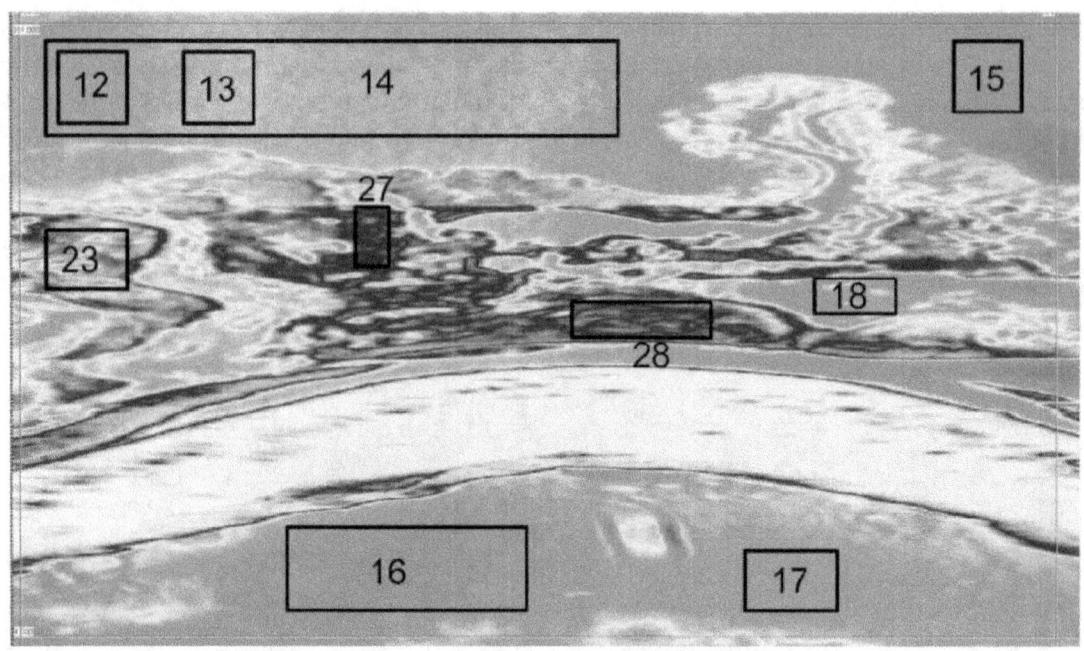

Figure 5.5 Interference Fit Data Image After First Brushing. The horizontal axis represents approximately −90 to +90 degrees. The vertical axis represents 360 mm (14.17 in.). Boxed regions were used in an amplitude analysis.

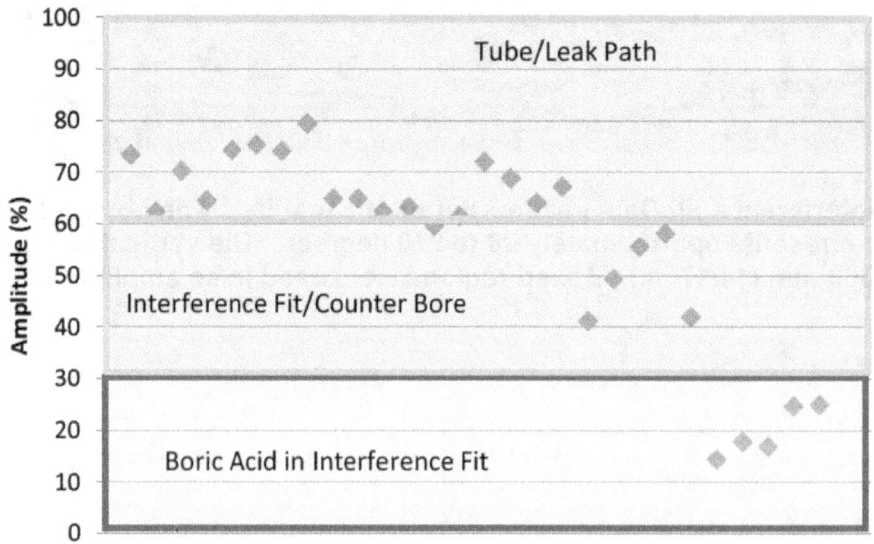

Figure 5.6 Mean Amplitude Response from the Regions Indicated in Figures 5.4 and 5.5

In summary, the mean amplitude responses from the interference fit of Nozzle 63 were measured and compared to the known responses for the different conditions of the interference fit of the mockup, as shown in Table 5.1. Note that the Nozzle 63 data were acquired with 1 dB more gain (13 dB as opposed to 12 dB) than the calibration mockup data and this difference was accounted for in the analysis.

Table 5.1 Mean Amplitude Responses (%)

Region	Calibration Mockup	Nozzle 63
Tube/Leak Path	60–75	60–79
Interference Fit/Counter Bore	40–55	41–58
Interference Fit – with Boric Acid	24–30	14–25

Based on the amplitude analyses conducted, the ultrasonic data for Nozzle 63 were also plotted with a tri-level color bar to represent the three interference fit conditions. The tri-level color bar implementation is shown in Figure 5.7. White represents the less than 30% amplitude range and indicates good transmission such as in the weld or possible boric acid in the interference fit region. Light blue represents the 30 to 60% amplitude range and indicates the nominal interference fit and counter bore regions. Dark blue represents the above 60% amplitude range and indicates poor transmission such as in the tube-to-air interface above the interference fit or an air gap or leak path in the interference fit and counter bore regions. A leakage path exists if a gap in the interference fit region extends fully through the interference fit connecting the weld and annulus region immediately above the weld to the dry side of the tube. From this image as well as the rainbow color-coded images, the ultrasonic data indicate one full leakage path, starting in the vicinity of 180 degrees circumferentially, or the low point of the nozzle, and meandering upwards and toward the right in the image. Other leakage paths are also evident but may not connect all the way through to the dry side of the assembly.

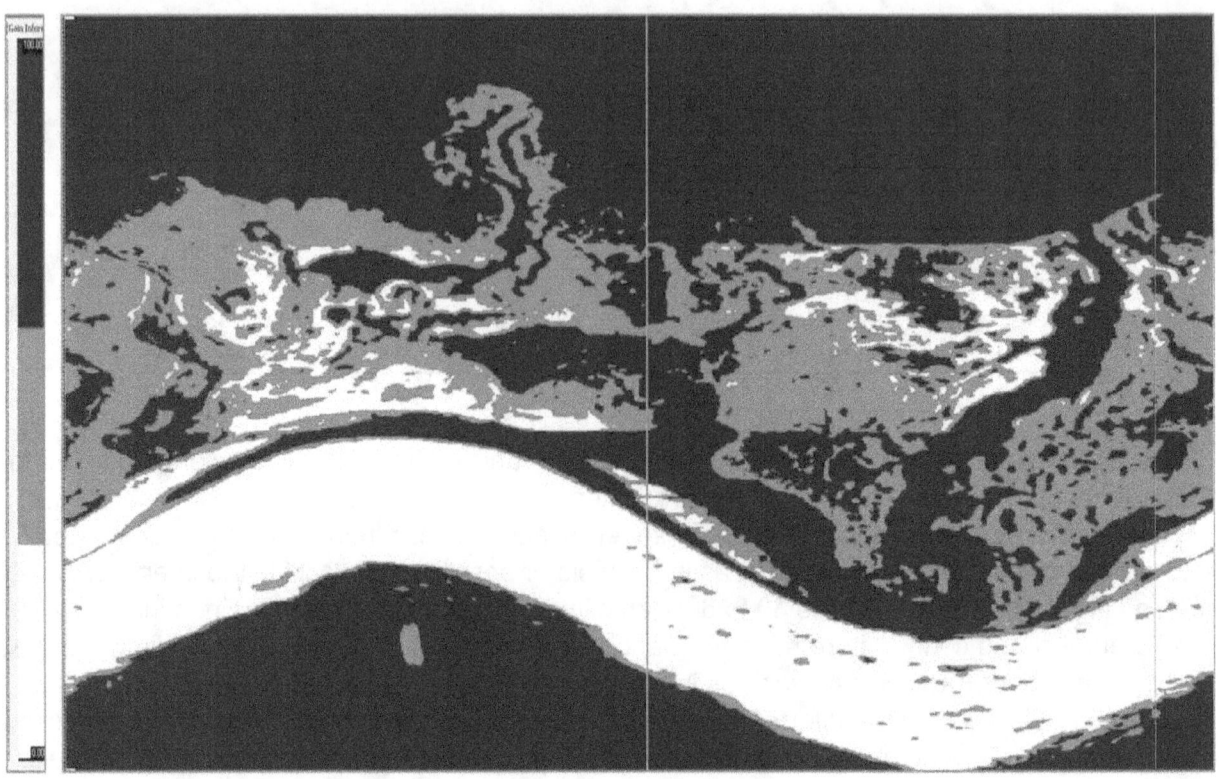

Figure 5.7   A Tri-Color Representation of the Interference Fit Data. The dark blue represents a leak path or tube-to-air interface, the light blue represents the nominal interference fit and counter bore regions and the white represents the weld and suspected boric acid presence in the interference fit.

5.4   Industry Standard Nondestructive Evaluations

Standard ultrasonic evaluation techniques used by in-service inspection (ISI) vendors include time-of-flight diffraction (TOFD) for detecting cracks in both the circumferential and axial orientation and zero-degree pulse echo for an interference fit examination. Blade probes (low profile) and solid-probe head configurations are deployed depending on the access conditions of the CRDM assembly (EPRI 2005; IAEA 2007, and discussions with JP Lareau, WesDyne International). An examination conducted by ISI vendor, WesDyne International (data supplied by JP Lareau), discovered a probable leak path in Nozzle 63 during a 2002 outage. The data acquired with industry-standard 5.0-and 2.25-MHz probes are shown in Figures 5.8 and 5.9, respectively. The 2.25-MHz image in Figure 5.9 has a lower resolution than the 5.0-MHz image in Figure 5.8 as expected, but both data sets detect the leak path observed at the low point (industry's zero-degree position) of the interference fit region.

The Figure 5.8 image can be compared to the PNNL's laboratory results in Figure 5.3. Both data sets were acquired with probes having nominal center frequencies of 5 MHz and they show the main leak path as a high-amplitude signal. In the WesDyne data, this is represented by the magenta color. Both data images capture other partial leak paths and show similar areas of high and low reflectivity. Although these field data sets were acquired in 2002 and a technique demonstration requirement did not come into effect until 2008, there have been only minor changes to the technique. The field procedure requires a minimum of a 51-mm (2-in.) scan above the weld, observation of a "river bed" pattern, and a nominal 2 dB difference in amplitude between the leak path response and the background.

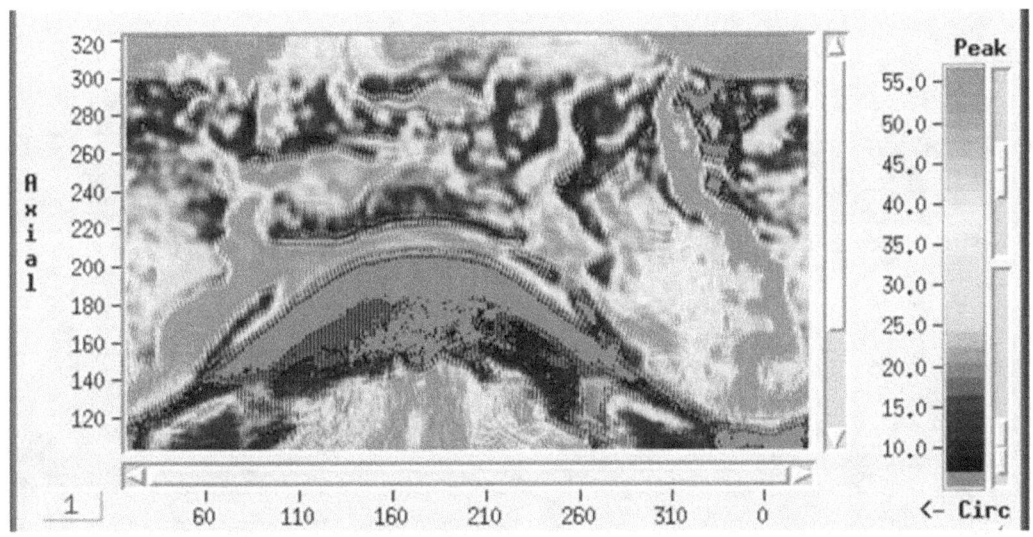

Figure 5.8   Ultrasonic Data from Nozzle 63 as Obtained by WesDyne International. The image was acquired with a 5-MHz probe. The horizontal axis represents the nozzle circumference in units of degrees. The vertical axis represents the nozzle axis in units of millimeters.

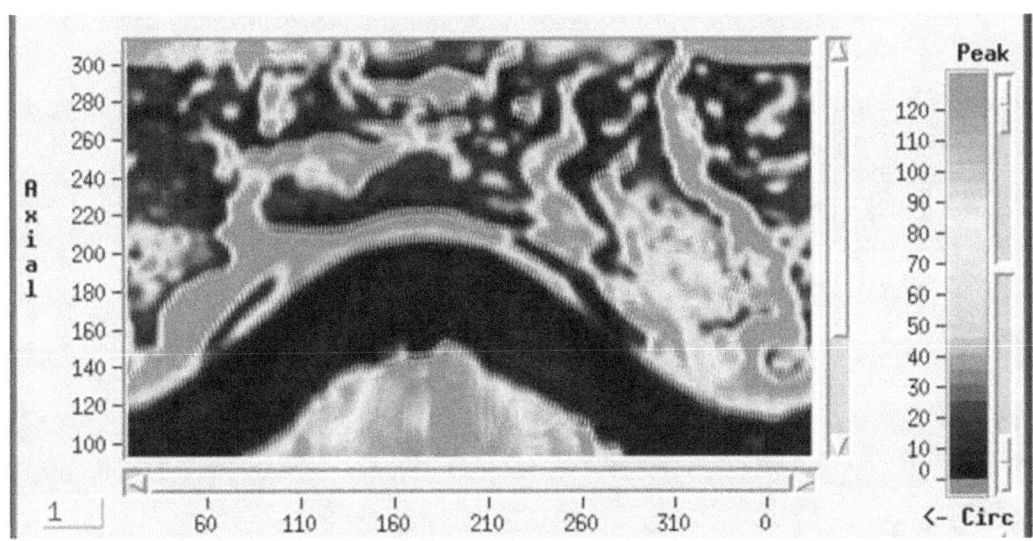

Figure 5.9   Ultrasonic Data from Nozzle 63 as Obtained by WesDyne International. The image was acquired with a 2.25-MHz probe. The horizontal axis represents the nozzle circumference in units of degrees. The vertical axis represents the nozzle axis in units of millimeters.

6 DESTRUCTIVE VALIDATION OF NOZZLE 63

To evaluate the efficacy of the ultrasonic characterizations of the leak path(s) and other areas of interest for Nozzle 63, as described in Section 5, the nozzle was destructively examined to allow visual or other characterization of the interference fit region. The destructive testing activity was conducted at the Babcock and Wilcox Technical Services Group (B&W) facility in Lynchburg, Virginia. This activity required removing the J-groove weld to separate the Alloy 600 penetration from the reactor pressure vessel (RPV) head. Pacific Northwest National Laboratory (PNNL) personnel were on site during the critical sectioning activities to identify proper cutting locations. Additionally, the J-groove weld region was preserved and returned to PNNL.

B&W used an industrial 5.79-m (19-ft) CobraFab Industries, Inc. band saw model VH2532HD with 'Cobra Strike' and an interchangeable blade option for all cutting purposes associated with the destructive activity. The initial size reduction cuts were performed on the Nozzle 63 specimen using one of the coarse toothed blades prior to the arrival of PNNL staff. Non-essential material was removed to reduce weight and to facilitate proper blade placement on the specimen during critical cuts. Figure 6.1 shows images from the size reduction activity. All cuts in this activity were conducted without lubrication or coolant to minimize disturbance to the interference fit region.

After size reduction, the nozzle assembly was prepared for the dissection cut that separated the high and low sides of the assembly. The cut line was selected to start at approximately the 95-degree mark (Figure 6.2), and follow through to the 275-degree mark. The line placement was based on the ultrasonic data and chosen to preserve the primary leak path previously identified in the ultrasonic images.

Figure 6.1 Size Reduction Cutting Activity

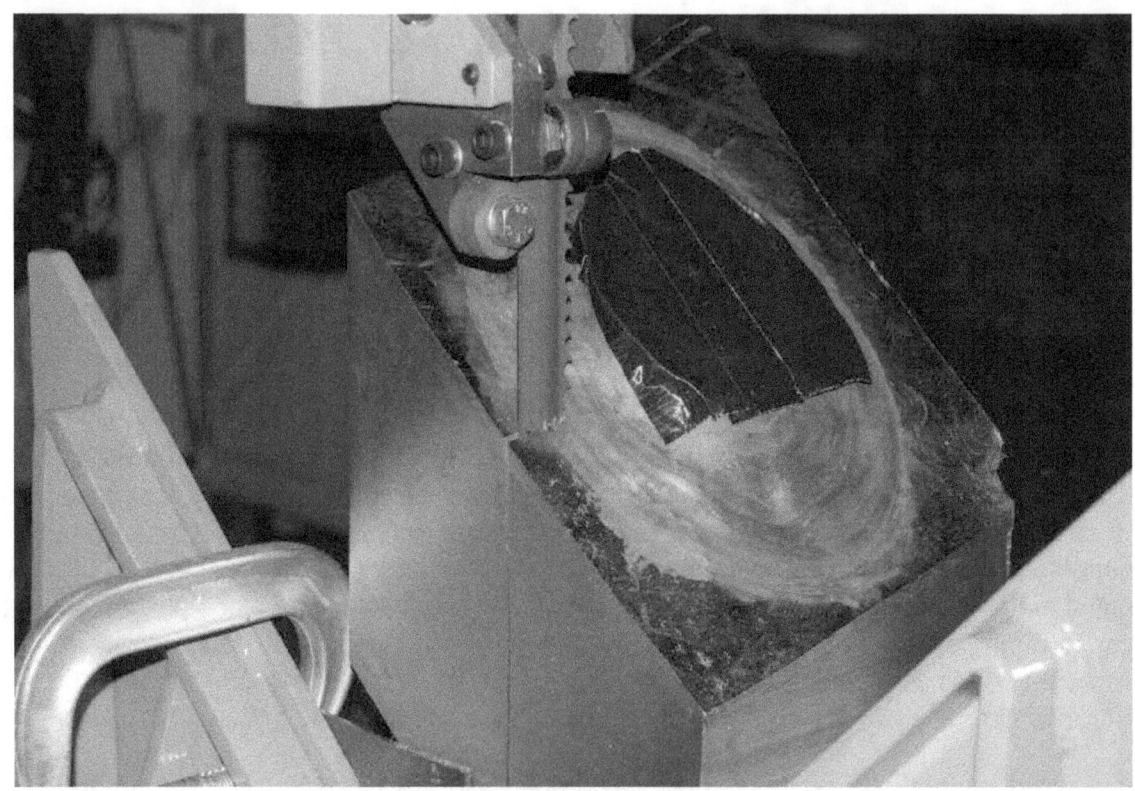

Figure 6.2 Start of the Dissection Cut

The cut to remove the J-groove weld was made 6.35 mm (0.25 in.) above the butter/triple point region in the RPV head as seen in Figure 6.3. The high or uphill side half-portion was selected for the first cutting. As previously stated, the 'high' side has several potential leak paths whereas the 'low' or downhill side had the primary leak path as identified in the ultrasonic data. The specimen was secured and the band saw tilted to an approximate 43-degree angle to match the angle between the nozzle and head. Cutting at this angle maximized the annulus region that was exposed while keeping the weld and butter regions intact for future evaluation. As the cut was designed to pass only through the RPV material, the fine toothed cutting blade was selected for use. However, other blades had to be used when cutting difficulties, as detailed in Appendix C, were encountered.

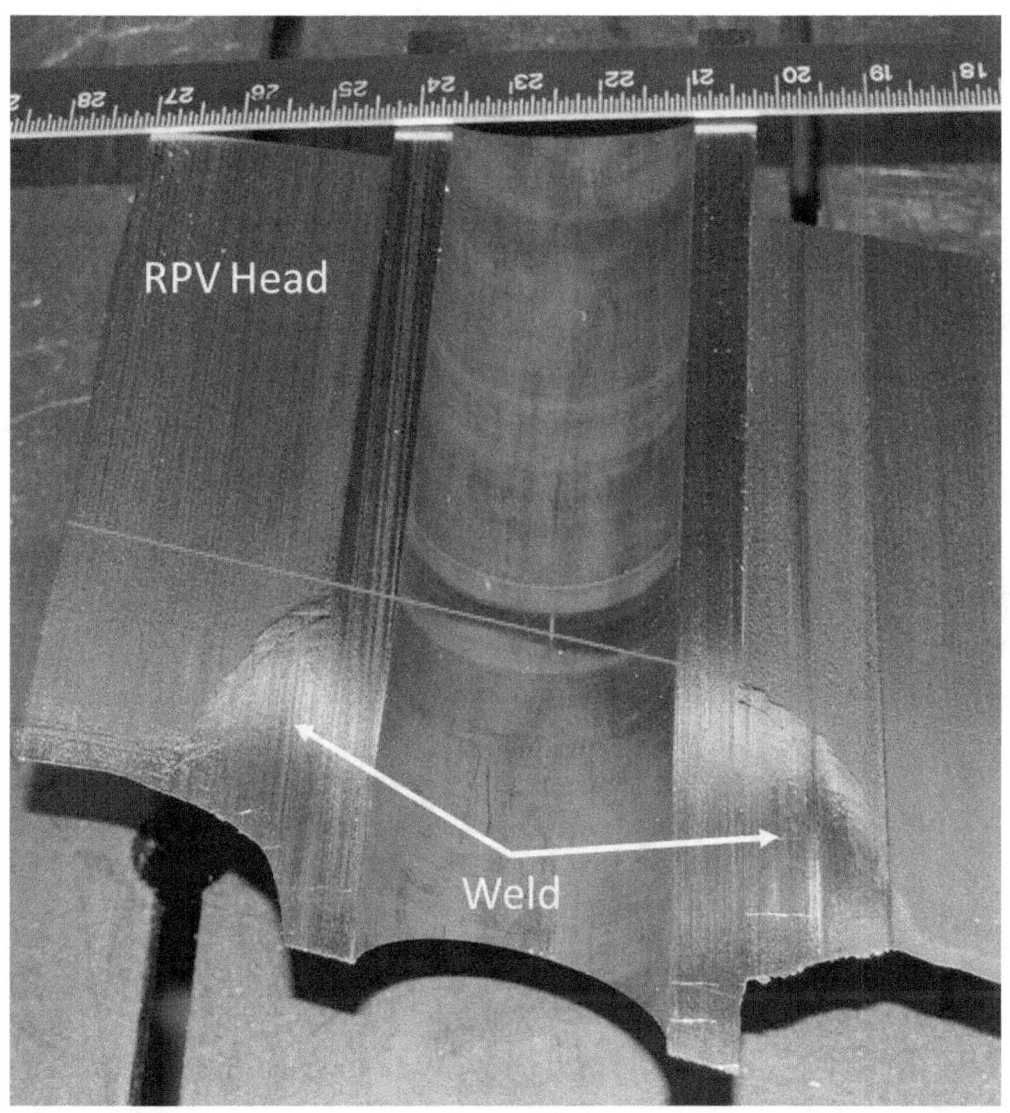

Figure 6.3   Nozzle 63 Assembly Cut in Half by Dissection Cut. The red line indicates where the weld removal cut was made.

The cutting continued until the entire nozzle was separated from the J-groove weld region as pictured in Figure 6.4. At this point, the nozzle region above the weld was freely released from the RPV head material. Removal of the nozzle exposed the annulus region of the high-side section as shown in Figure 6.5. A Nikon D40x camera was used to acquire high-resolution photographic documentation of the annulus region. A subsequent cut was conducted on the low-side section to expose its annulus region containing the primary leak path (Figure 6.6). The nozzle freely released from this portion as well.

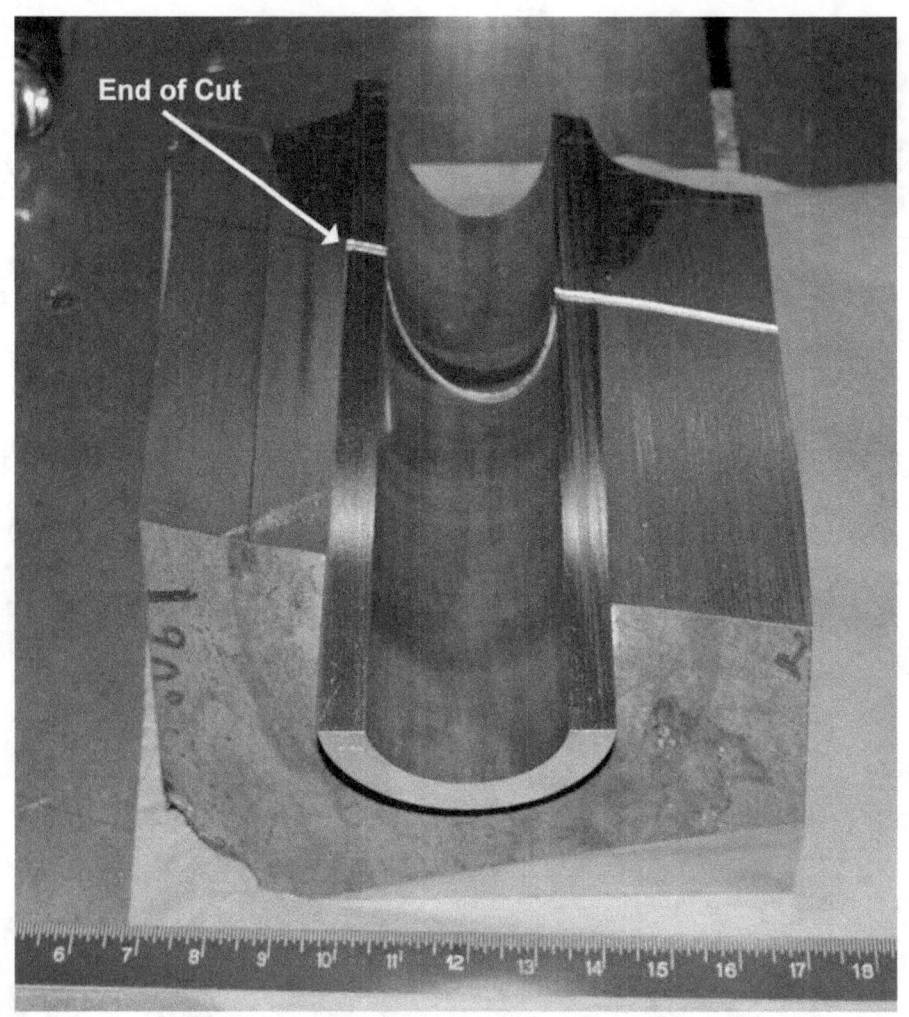

Figure 6.4 End of J-groove Weld Removal Cut

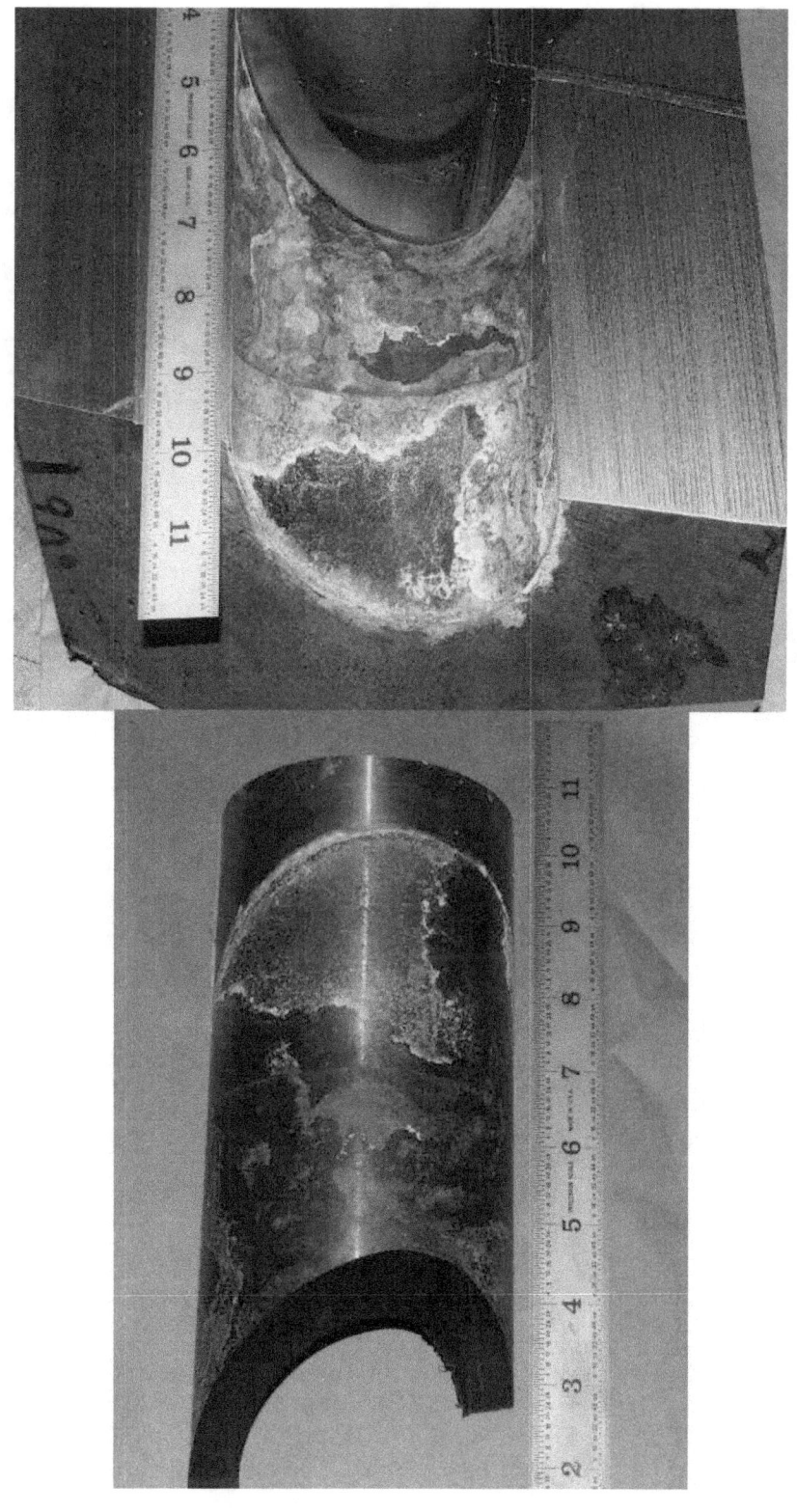

Figure 6.5 Exposed RPV Head and Nozzle from High Side Section

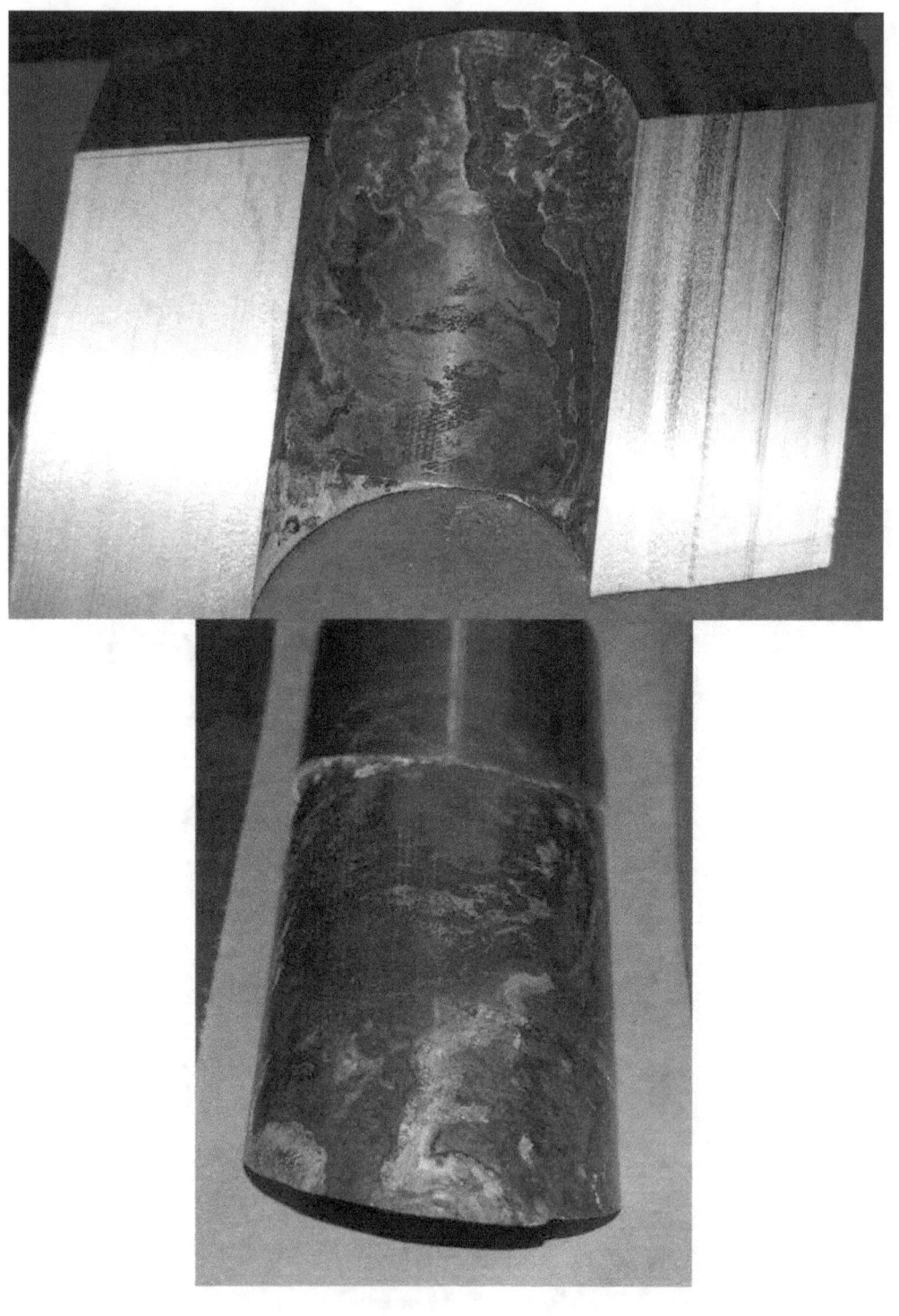

Figure 6.6 Exposed RPV Head and Nozzle from Low-Side Section

7 CORRELATION OF ULTRASONIC AND DESTRUCTIVE RESULTS

This section compares the phased-array ultrasonic results to the visual results obtained by cutting through the nozzle assembly to reveal the interference fit surfaces. The nozzle outer diameter (OD) surface was photographed in 45-degree increments with the individual photographs cropped and stitched together to form the montage image in Figure 7.1. Some evidence of thin boric acid deposits is visible in the white regions while a thin corrosion layer is seen in the rust-colored regions. The red line marks the interference fit region. The main leak path is identified by the two black arrows.

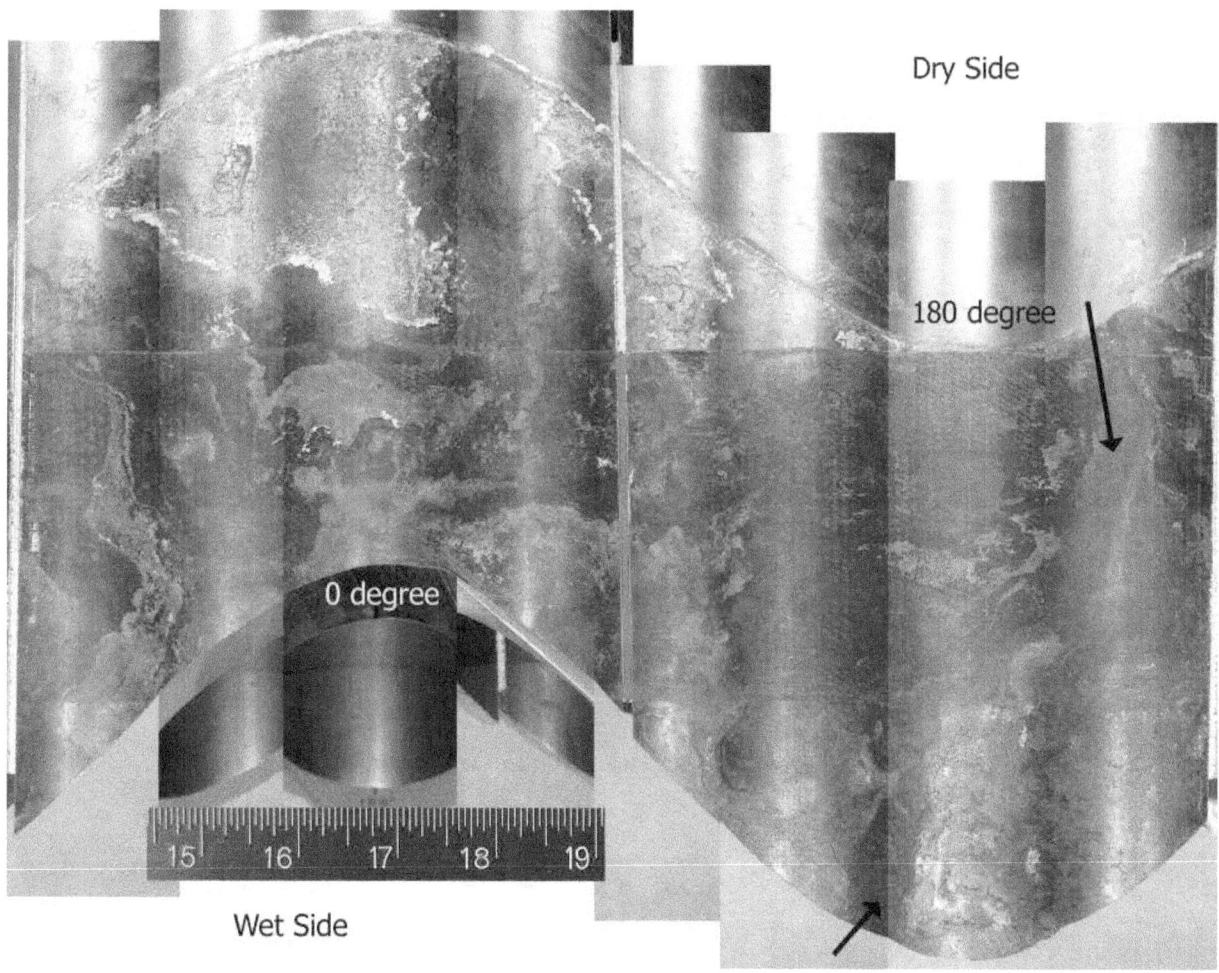

Figure 7.1   Nozzle Surface. The red line marks the interference fit region and the two black arrows identify the main leak path.

Similarly, the exposed reactor pressure vessel (RPV) head was photographed and the stitched image is displayed in Figure 7.2. The main leak path and other features seen in the ultrasonic images are clearly evident. Boric acid deposits are visible in white and corrosion products in the rust color. The interference fit region is evident in the photograph and is marked with the red line. For comparison, the ultrasonic data image with the same aspect ratio as the photograph is displayed in Figure 7.3. The ultrasonic image was stretched to best fit the visual data but the match is not perfect due to the curved surfaces. Nevertheless, the ultrasonic features well match the features seen visually on the RPV head annulus. Clearly, the main leak path and other partial leak paths are evident and well imaged.

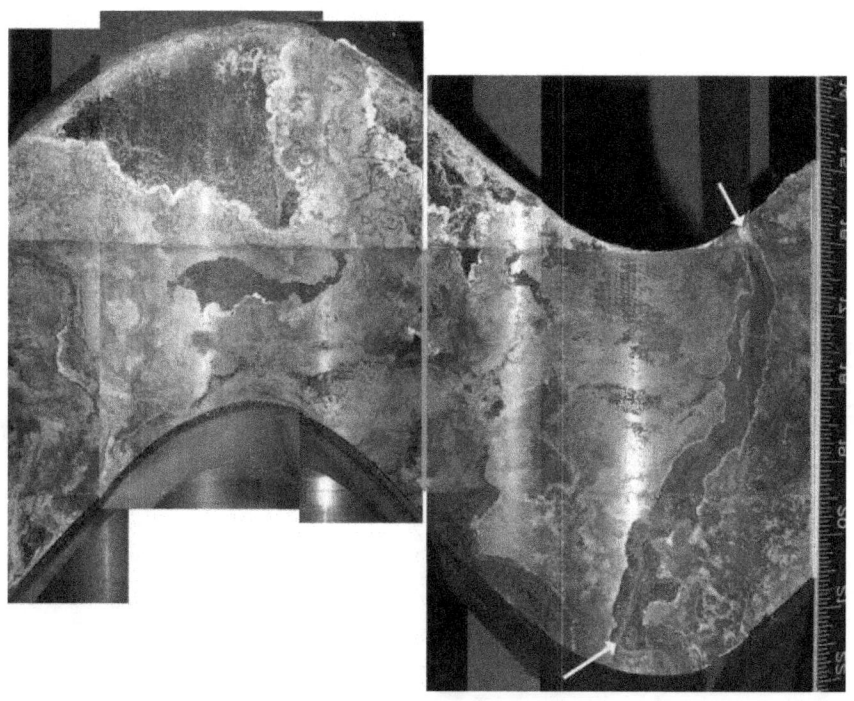

Figure 7.2   RPV Head Surface. The red line marks the span of the interference fit region and the two yellow arrows identify the main leak path.

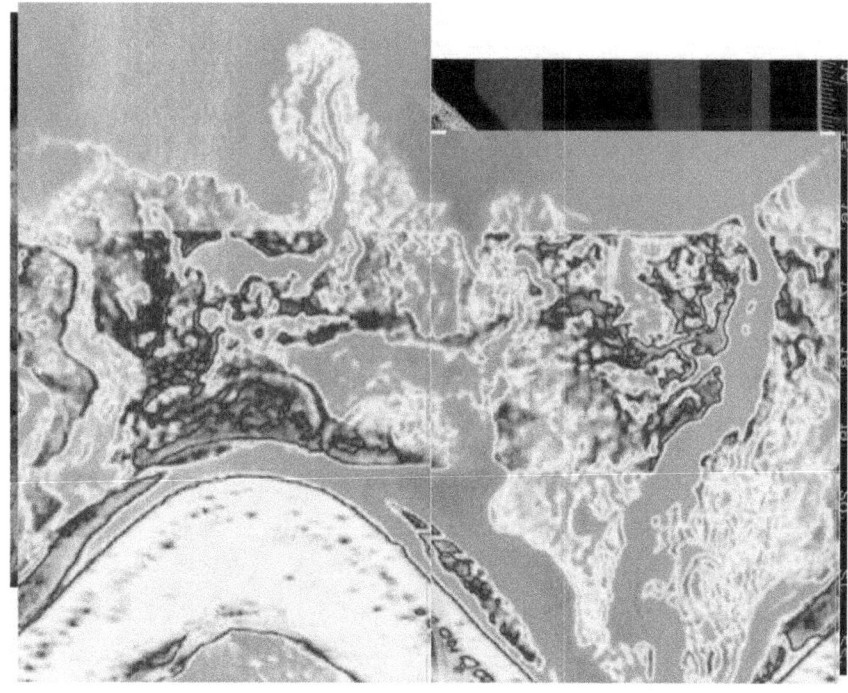

Figure 7.3   Ultrasonic Data Stretched to Best Match the RPV Head Photograph

8 ADDITIONAL PHYSICAL MEASUREMENTS ON THE REACTOR PRESSURE VESSEL HEAD

The visual evaluation of the exposed interference fit region showed no indication of cracking or significant corrosion in the annulus of the nozzle assembly pieces. During plant operation, it is expected that the annulus will open up between the penetration and RPV head. If reactor coolant is allowed into the annulus through a crack, it could spread throughout the entire annulus region by a path or paths of least resistance. This infiltration of borated water into the interference fit over time results in boric acid deposits and possibly corrosion of the RPV head. Degradation of the Alloy 600 nozzle could occur from steam cutting of the nozzle surface during plant operation and opening of the annulus. Evidence of this type of degradation was not found on the Nozzle 63 surface.

Following the visual comparison of the interference fit region to the ultrasonic data, attempts were made at measuring the boric acid thickness in the annulus as well as the extent of corrosion of the reactor pressure vessel (RPV) head to see how these conditions affect the ultrasonic signal response. The boric acid thicknesses were first measured at specific points using a calibrated eddy current thickness gage. Next the RPV head surface was replicated with a Microset material, and boric acid thickness measurements were made on cross-sectional slices through the replica in the main leak path area. Finally, the replicated sections were examined with a stereomicroscope providing an indication of the corrosion extent. These activities are described in this section.

8.1 Boric Acid Deposit Thickness Measurements

In addition to visually comparing the condition of the interference fit region to the ultrasonic data, the thickness of boric acid deposits on the RPV head in the annulus region was measured to see how this correlated to the ultrasonic signal response. It is recognized that the boric acid deposit thickness would not be the only factor to affect the ultrasonic signal response. Other factors could include boric acid deposits on the nozzle material (refer to Figure 7.1), the density of the boric acid deposits, and the difference between boric acid and the visible low-alloy steel corrosion product. Nevertheless, the acquired boric acid thickness measurements should provide some insight for interpreting the ultrasonic data.

A DeFelsko PosiTector 6000 Series eddy current coating thickness gage was used to measure the boric acid deposit thickness at selected points in the annulus region on the RPV head material. The probe had a point contact area of 1 mm (0.039 in.) in diameter and spanned a measureable coating thickness range of 0 to 1.14 mm (0 to 45 mils). The thickness gage accuracy was verified by taking measurements on several plastic shims of known thickness in the 24 to 507 micron (0.94 to 20.0 mils) range. Measurement error was less than or equal to 2.5 microns (0.098 mils). The selected measurement sites were chosen to represent the differing ultrasonic amplitudes corresponding to the various conditions in the interference fit region, including the main leak path, other partial leak paths, the interference fit with and without suspected boric acid deposits present, and areas outside of the interference fit. A total of 70 measurement points were selected by both Pacific Northwest National Laboratory (PNNL) and U.S. Nuclear Regulatory Commission (NRC) personnel and are displayed in Figure 8.1 for the photographed uphill and downhill halves of the RPV head. The red dots represent what appear to be leak paths or bare metal regions. The green dots represent areas with different reflectivity in the photo and differing ultrasonic response. Lastly, the yellow dots represented additional points of interest including two pairs of points on either side of the upper interference fit border.

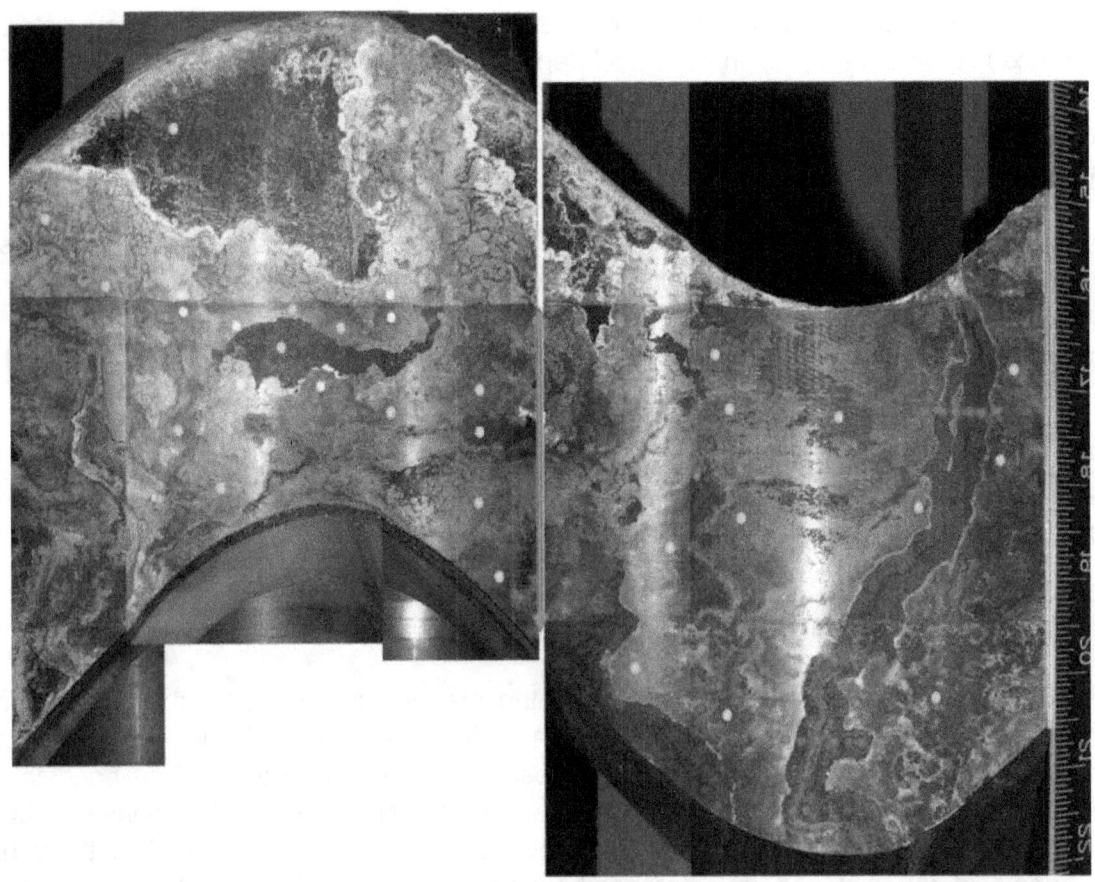

Figure 8.1   Photograph of the RPV Head Material with Boric Acid Measurement Points. The interference fit region is marked by the red line.

Each of the 70 data points was numbered and the boric acid measurements were entered into a spreadsheet. The thickness measurements in microns are displayed adjacent to the data points in Figure 8.2. Some general observations were:

1) Where boric acid deposits are present above and below the interference fit region, their thicknesses are nominally in the 130- to 200-micron (5.1- to 7.9-mils) range.

2) The two pairs of data (yellow dots) on either side of the interference fit on the uphill section show boric acid deposit thicknesses of 156 and 150 microns (6.1 and 5.9 mils) above the fit region and 62 and 74.5 micron (2.4 and 2.9 mils) in the fit region. This suggests that the

deposit thickness may be less within the span of the interference fit than within the annulus above or below.

3) The leak path and what appear as bare metal or nearly bare metal locations have a thin surface layer of corrosion deposits rather than visible boric acid deposits, with corrosion deposit thicknesses of 16 microns (0.63 mils) or less. These are thinner than the measured boric acid deposits inside or outside the span of the interference fit.

The relationship between the ultrasonic response values and eddy current (EC) thickness values was explored in more detail. As demonstrated in the mockup evaluation, the large-amplitude ultrasonic data represents an air gap in the interference fit between the penetration and the RPV head. These data regions are represented by the color orange in the rainbow color bar previously shown in Figure 5.3, for example. Two horizontal profiles that display ultrasonic amplitude variations along the selected measurement line in the image are shown in Figure 8.3. The left side displays a profile taken along the red horizontal line located in the interference fit region marked on the ultrasonic image. Clearly the high ultrasonic response is evident across the main leak path and then the response falls off on either side, of the leak path. The left side drop off is more severe in the profile and is associated with the blue color in the image, while the right side drop off is moderate and is associated with the yellow color in the image. The right side displays a profile taken along the red horizontal line in the ultrasonic image below the interference fit. The profile as well as the data image, again, shows a high-amplitude ultrasonic response for the leak path regions with signal drop off on both sides of the leak paths. At this axial position in the annulus, the drop off is smaller than that observed in the data from within the interference fit on the left. For comparison, the exposed interference fit region at the low-alloy steel surface is displayed in the lower portion of the figure with the blue lines corresponding to the profile locations.

The inherent characteristic of a leak path ultrasonic response, as confirmed by visual examination of the interference fit, is a high-amplitude response that vertically traverses the entire interference fit (I fit) region. The primary leak path deposit layer measurements obtained with the eddy current probe further show that within the leak path there will be an absence of deposits and therefore an air gap between the nozzle penetration and RPV head. This inverse relationship between the deposit thickness and the amplitude of the ultrasonic response is represented in Figures 8.4 and 8.5. Eddy current layer thickness data obtained at the red points horizontally spanning the primary leak path in Figure 8.1 are plotted as boric acid (BA) data in blue diamonds, realizing that these are boric acid and corrosion layer values. The corresponding ultrasonic data at those same points are plotted as ultrasonic test (UT) data in red squares. Figure 8.4 represents data obtained in the interference fit region and Figure 8.5 represents data outside of the interference fit from the primary leak path points, red in Figure 8.1. In each figure below, the leak path boundary is noted by two black horizontal lines. Both figures show that leak path data points are represented by this characteristic high ultrasonic response (above 70%) and low boric acid or corrosion layer thickness (approximately below 20 microns [0.79 mils]). More specifically, the BA or corrosion layer thickness within the interference fit is below 7 microns (0.28 mils) with the exception of the data point acquired at the island position in the main leak path. Here a larger thickness value of 25.5 microns (1.0 mils) was measured. Similarly, for data outside of the interference fit but still in the primary leak path, the layer thickness values were below 12 microns (0.47 mils) except for the data point at the

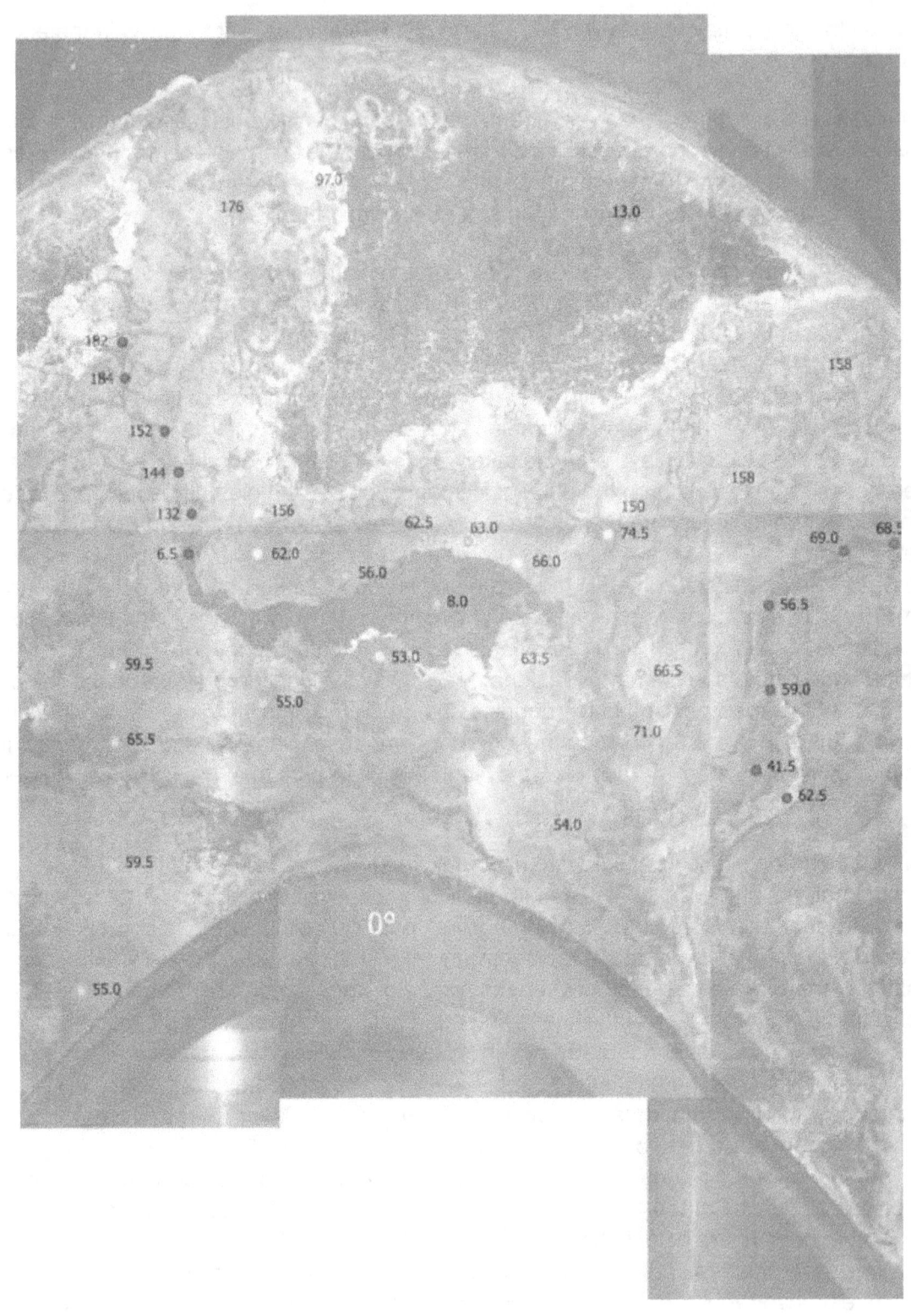

Figure 8.2 Boric Acid Deposit Thickness Values in Microns

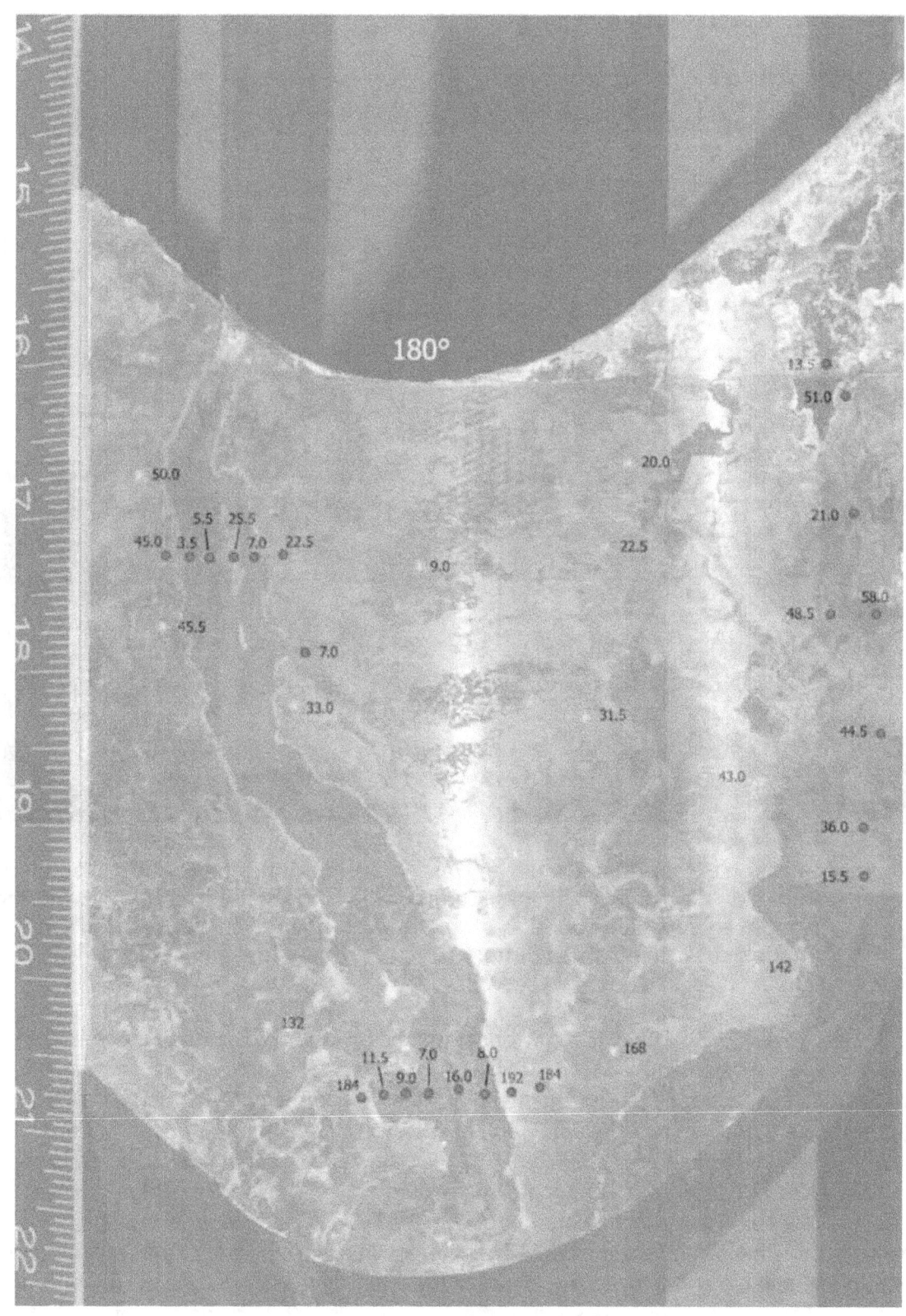

Figure 8.2 (Continued)

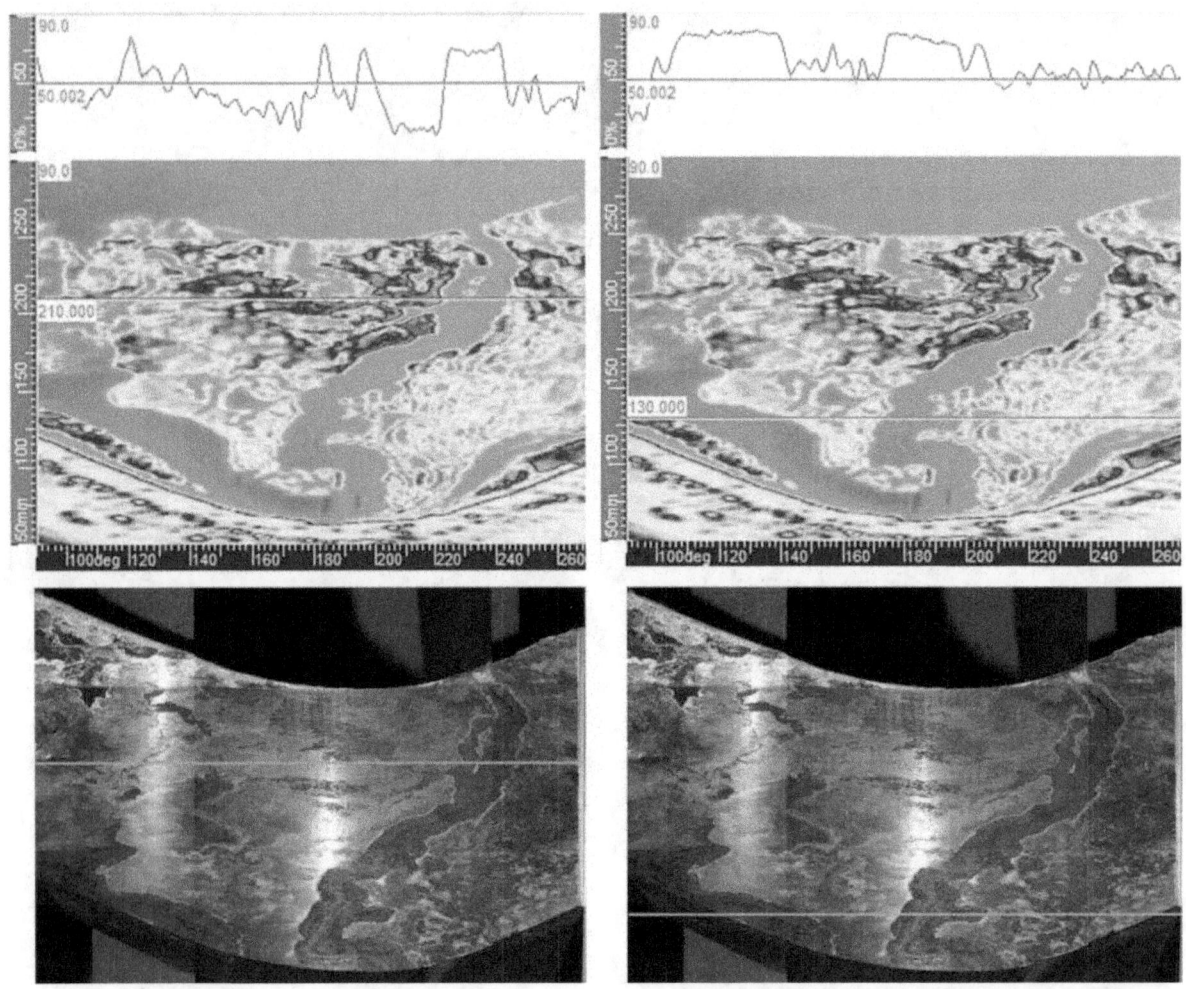

Figure 8.3 Ultrasonic Response Profile at Top, Acquired Along the Red Horizontal Line in the Middle Image. The left image profile is in the interference fit and the right image profile is outside of the interference fit. Lower photographic images show the exposed interference fit on the low-alloy steel surface.

rust-colored streak in the middle at 16 microns (0.63 mils). Moving outside of the leak path, the ultrasonic response drops and the BA/corrosion layer increases. Within the interference fit, Figure 8.4, this increase in BA/corrosion layer thickness is much smaller as can be expected. The interference fit region has little to no gap between the nozzle and RPV head but this condition is variable throughout the fit (Hunt and Fleming 2002). Boric acid presence in this region would likely be a thin layer and more compacted as noted by the low (blue-colored) ultrasonic response. Outside of the interference fit, the BA/corrosion layer increases more dramatically on either side of the main leak path, to nominaly 185 microns (7.28 mils). A wider gap was present in the annulus region outside of the interference fit due to a machined counter bore (refer to Figure 1.1). The data suggest that this gap fills with boric acid. Note that the ultrasonic response shows a smaller drop on either side of the leak path possibly indicating that the boric acid was less compacted in this region.

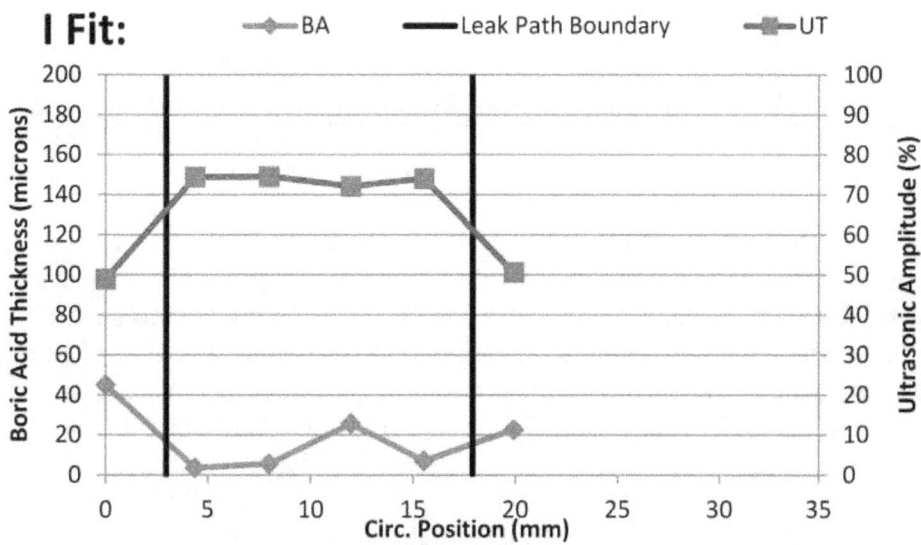

Figure 8.4 Ultrasonic and Boric Acid/Corrosion Layer Thickness Along a Line Across the Main Leak Path in the Interference (I) Fit

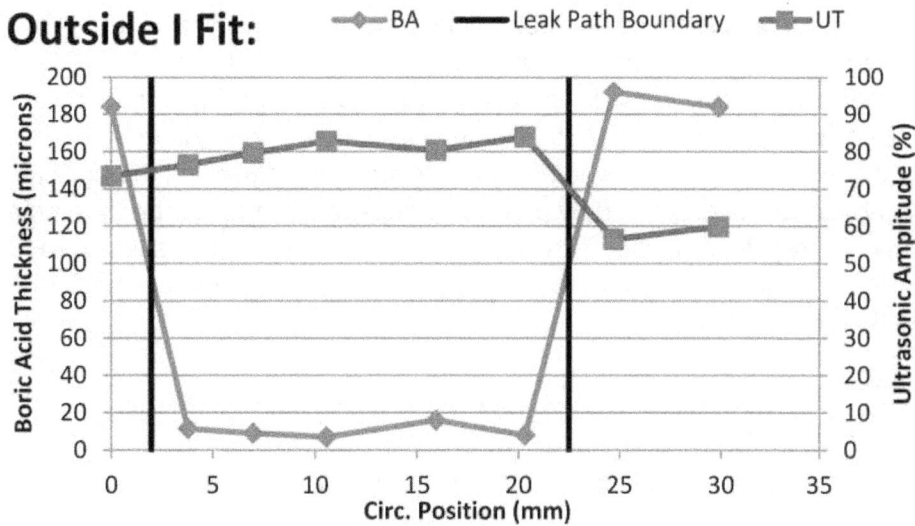

Figure 8.5 Ultrasonic and Boric Acid/Corrosion Layer Thickness Along a Line Across the Main Leak Path Outside the Interference Fit

Lastly, an attempt was made to correlate the ultrasonic and eddy current measurements at all 70 data points. In this analysis, the data points were segregated as within the interference fit region or outside of the interference fit region. Initially, all of the points were plotted with results shown in Figure 8.6. Points with high ultrasonic response and low BA or corrosion deposit thickness, upper left, are primarily associated with the main leak path region and represent areas where any potential deposits are flushed out and not allowed to remain. In addition to the primary leak path, other partial leak paths and regions that visually appear as bare metal were identified. All of these leak path and bare metal points are plotted in Figure 8.7. These points represent a likely gap in the annulus between the nozzle penetration and the RPV head, noted by the high ultrasonic response. The primary leak path points from both within and without the interference fit are represented in the upper left portion of the graph as exhibiting high ultrasonic responses and minimal layer deposit thicknesses. The interference fit data in the middle, represented by approximately a 40- to 70-micron (1.58- to 2.76-mils)-thick deposit and 45 to 75 percent UT response, corresponds to the possible secondary leak path points. Finally, the data outside the interference fit in the upper right portion of the graph represent the points in the plume above the interference fit. These points are located on a suspected secondary leak path that exits the interference fit and extends upward into the annulus. The ultrasonic responses along this "plume leak path" are high, 55 to 83 percent. The boric acid values are also high, 132 to 182 microns (5.20 to 7.17 mils). It is likely that the boric acid was able to collect in this upper annulus flow channel due to the wider gap in the presence of a counter bore and an assumed low flow rate. The remaining points (70 original points except for the leak path and bare metal points) are plotted in Figure 8.8. An ultrasonic amplitude level of approximately 50 percent divides the data, with lesser values taken from locations within the interference fit and greater values taken from locations outside the interference fit region. For this nozzle, the BA or corrosion deposit thickness values in the interference fit are less than 75 microns (3.0 mils), and outside the interference fit are greater than 130 microns (5.12 mils). The outlier point at the center of the graph, with a layer thickness of 97 microns (3.82 mils) in Figure 8.8, was obtained at a location where there is a noticeable boric acid deposit on the nozzle side. The thickness of this deposit, though not quantitatively known, would increase the layer thickness value moving the point to the right and likely in the range of the other "Outside I Fit" data points. The other outlier point (EC value of 9 microns [0.35 mils] and UT value of 73%) is from the interference fit data and was obtained at a rust-colored region so it was not included in the bare metal points. This outlier is also possibly explained by positional registration error of the data points. Nevertheless, data trends are evident. While the numerical values are specific to Nozzle 63, these trends are expected to be applicable for upper head penetrations in general. A summary of the Nozzle 63 results follows.

- Leak paths
    a. High UT response, >60%
    b. Minimal boric acid/corrosion layer EC measurement, <16 microns (0.63 mils)
- Outside the leak path
    a. Inside the interference fit region
        i. UT response <50%
        ii. EC measurement low, 15–75 microns (0.59–3.0 mils)
    b. Outside the interference fit region
        i. UT response >50%
        ii. EC measurement larger, 130–190 microns (5.12–7.48 mils)

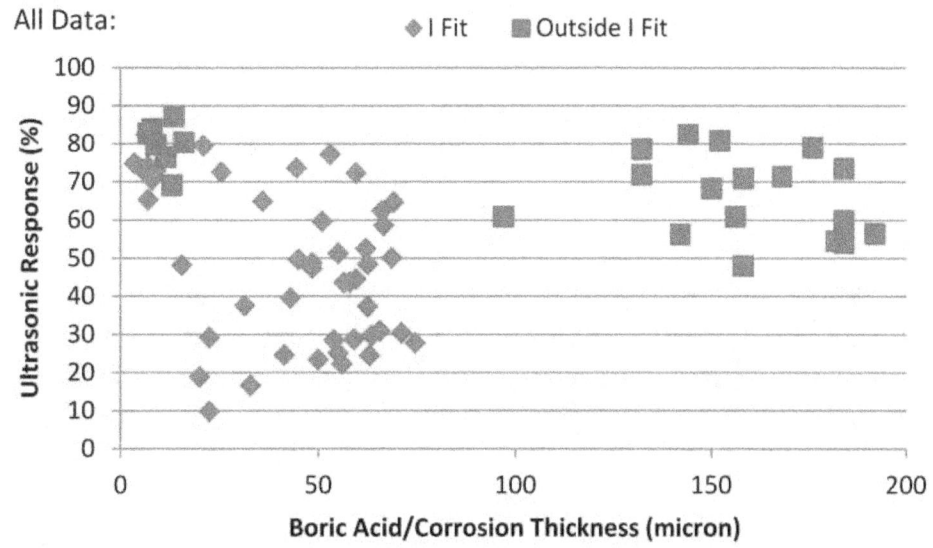

Figure 8.6 Comparison of Ultrasonic Response and Boric Acid/Corrosion Layer Thickness at all 70 Acquired Data Points

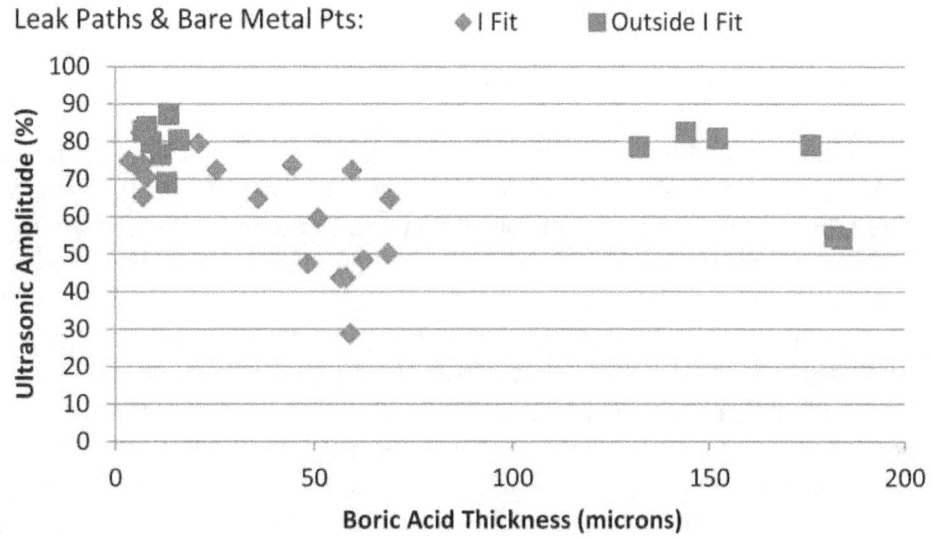

Figure 8.7 Comparison of Ultrasonic Response and Boric Acid/Corrosion Layer Thickness at the Leak Path and Bare Metal Data Points

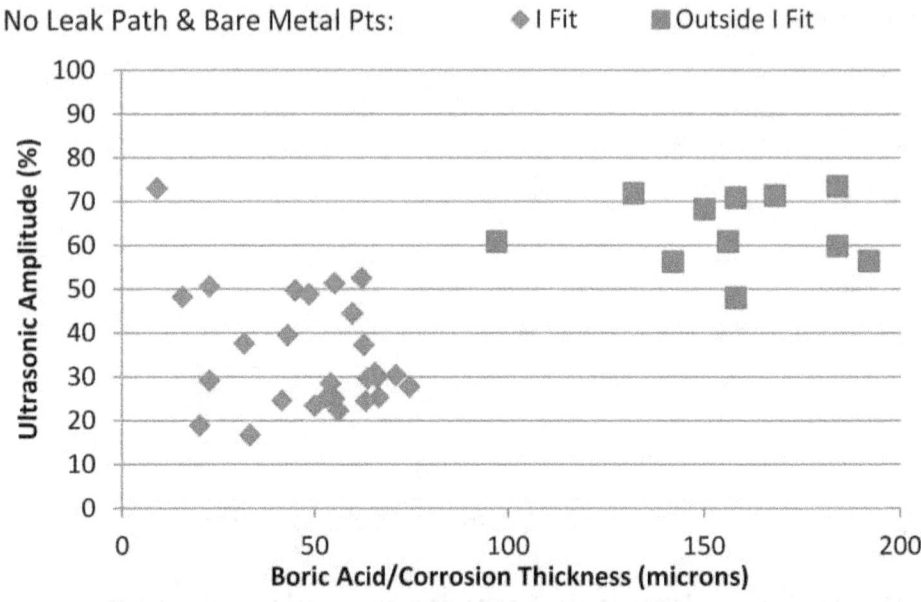

Figure 8.8   Comparison of Ultrasonic Response and Boric Acid/Corrosion Layer Thickness at Data Points Not in the Leak Paths Nor at Bare Metal Points

8.2   Leak Path Surface Replication

Following the surface deposit thickness measurements with the eddy current probe, the profile of the main leakage path on the RPV head side of the annulus was analyzed by surface replication. The purposes for the surface replication were to compare the surface deposit thicknesses to those measured by the EC probe and to identify any corrosion or loss of material from the RPV surface. This replicating material has better than 0.1-micron (0.004-mils) resolution. Eight areas of interest across the main leak path, as shown by the horizontal lines in Figure 8.9, were selected where the replicant would be cut for cross sectional analysis. The sliced and annotated replica is shown in Figure 8.10. The surface deposit thickness was measured from the cross section pieces at the locations indicated by the white arrows in Figure 8.10. Several of the measurement values obtained by these surface cross sections were compared to the thickness gage measurements described in Section 8.1. The root mean squared error for the respective measurement methodologies was 14 microns. This error is assumed to be primarily due to data registration.

To evaluate the surface condition of the leakage path for evidence of loss of material from corrosion, the replicated surfaces from Figure 8.10 were viewed with a stereomicroscope. As shown in Figures 8.11 and 8.12, machining marks were observed on the replicated surfaces indicating minimal corrosion, erosion, or wastage throughout the leak path region, likely indicating a relatively low leakage flow rate during plant operation. Figure 8.11 shows replica pieces 2 and 3 in the main leak path in the region below the interference fit. Both pieces show double streaks from corrosion product staining but no or minimal actual corrosion or wastage. The machining marks are intact across the images. The transition from below the interference

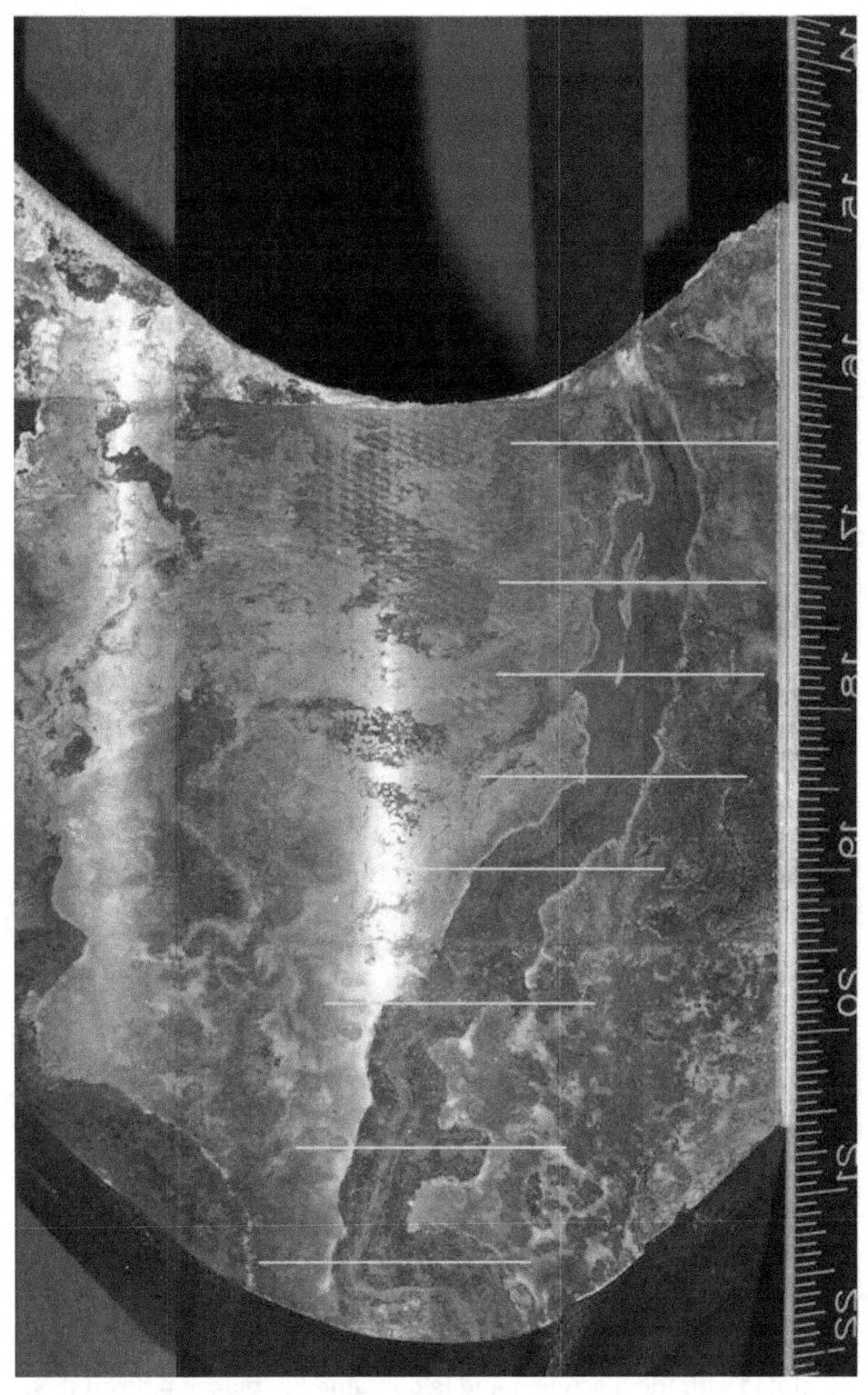

Figure 8.9   Eight Areas Selected for Boric Acid Thickness Measurements on Cross-Sectional Slices of Microset Replica

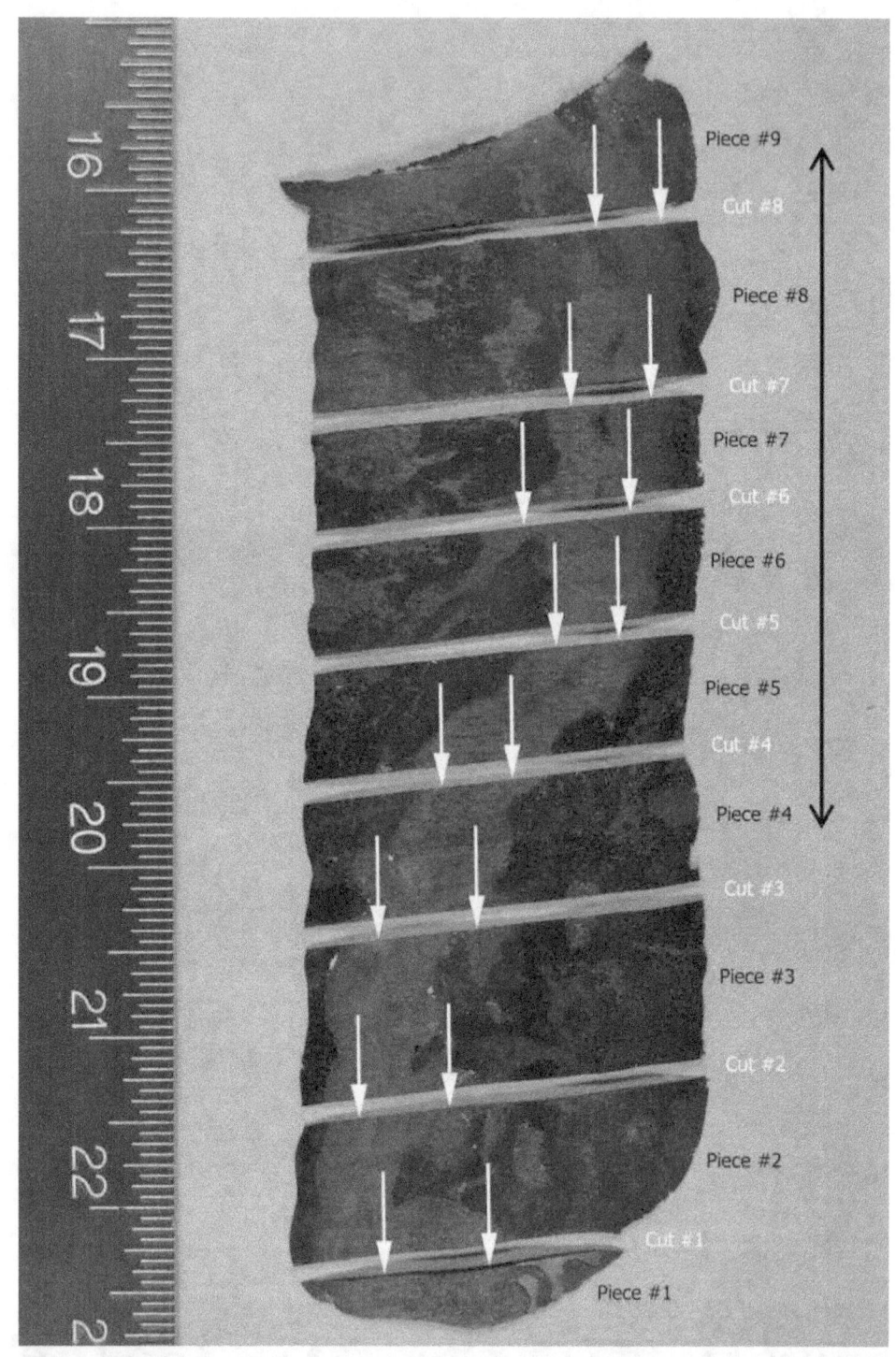

Figure 8.10 Leak Path Replica with Cuts and Pieces Identified. The interference fit region is noted with the black line and is contained in pieces 4 through 9.

Figure 8.11 Staining Streaks in the Leak Path Below the Interference Fit from Replica Pieces 2 and 3, Left and Right, Respectively. The red line represents 2.0 mm (0.080 in.) in length.

Figure 8.12 Transition from Below the Interference Fit to the Interference Fit Region. Machining marks are evident in this replica piece 4. The red line represents 2.0 mm (0.080 in.) in length.

fit to the interference fit region is captured in Figure 8.12 on piece 4. Machining marks are clearly evident and were observed in most of the bare areas examined on the RPV head surface. The surface finish within the interference fit region was approximately equivalent to a turned finish of 1.6 micro-meters (63 microinches). The finish below the interference fit region was approximately equivalent to a milled 1.6 micro-meters- (63 microinches-) finish. Piece 5 contained an angular feature or anomaly with an approximate length of 2.3 mm (0.090 in.) and is shown in Figure 8.13. The right image is at a twice the magnification as the image on the left and shows more detail. This feature appeared to be more of a dent or scrape and not corrosion. Piece 5 lies in the interference fit region. The only corrosion observed in the replicated surfaces was in the region above the interference fit in piece 9. The piece is shown in Figure 8.14 with the two areas of interest circled. The circled region on the left was in the main leak path and covered an area of approximately 6.4 mm (0.25 in.) in diameter with a depth of 0.25 mm (0.01 in.). The corroded area on the right was approximately 12.7 by 1.6 mm (0.5 by 0.06 in.) with a depth of 0.25 mm (0.01 in.).

 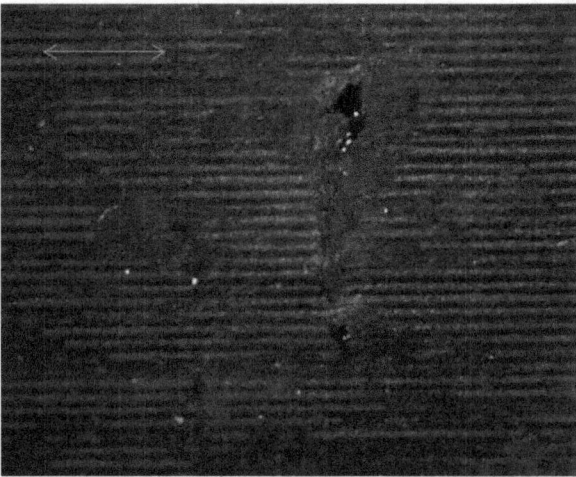

Figure 8.13 Piece 5 from the Interference Fit Region Shows an Indication of a Scrape. In the left image the red line represents 2.0 mm (0.080 in.) in length. The image on the right at twice the magnification of the left shows more detail.

Figure 8.14 Corrosion Areas Observed Above the Interference Fit Region. The red line represents 2.0 mm (0.080 in.) in length.

9 CONCLUSIONS

1. The purpose of this investigation was to evaluate the efficacy of ultrasonic testing (UT) for detecting leakage paths for reactor coolant through the interference fit between upper reactor pressure vessel (RPV) head penetrations and the RPV head. Leaking could occur from primary water stress corrosion cracking of the J-groove weld in the nozzle assembly. The evaluation was undertaken by acquiring UT data from a suspected leaking penetration removed from plant service, and confirming the features such as the leak path and boric acid deposits by destructive visual examination. The subject for the investigation was control rod drive mechanism (CRDM) Nozzle 63 from the North Anna Unit 2 plant, which was suspected to be leaking when the RPV head was replaced in 2002.

2. Prior to testing Nozzle 63, the phased-array UT system was tested on a CRDM mockup. Testing on the mockup indicated that the system could distinguish various conditions in the interference fit. A nominal interference fit was indicated by a mid-amplitude ultrasonic response. Boric acid deposits in the interference fit were identified by a lower amplitude response relative to the nominal fit, and an air gap in the interference fit could be identified by a higher amplitude response relative to the nominal fit.

3. The phased-array ultrasonic evaluation of Nozzle 63 indicated a potential primary leakage path on the downhill side of the nozzle, as characterized by a high-amplitude ultrasonic response that would indicate a gap between the nozzle penetration and RPV head. Scattered boric acid deposits were also indicated throughout the annulus region by a relatively low-amplitude response.

4. Following acquisition of the UT data, Nozzle 63 was destructively examined by segmenting to expose the RPV head and nozzle penetration surfaces in the interference fit region. Visual examination confirmed the presence of the leakage path and boric acid deposits in the locations indicated in the ultrasonic data.

5. Eddy current measurements of boric acid and corrosion surface-deposit thickness on the RPV head surface of the interference fit region showed a rapid transition from high thickness just outside the leak path, to low thickness in the leak path (<16 microns [0.63 mils]), and back to high thickness on the other side of the leak path. This corresponded to a rapid change in the ultrasonic response from low amplitude, to high (>60%), and low again. Both the ultrasonic and eddy current responses appear to be characteristic of a well-defined leakage path.

6. Generally, the region of the interference fit gives a lower ultrasonic amplitude response (<50%) compared to regions outside the interference fit (>50%) suggesting a relatively more compact boric acid or corrosion deposit layer that efficiently transfers ultrasonic energy.

7. The thickness of boric acid or corrosion deposits is generally measured to be higher outside the interference fit (130 to 190 microns [5.12 to 7.48 mils]) compared to inside the interference fit (15 to 75 microns [0.59 to 3.0 mils]). This is expected due to the wider gap in the annulus region in the presence of a counter bore.

8. Surface replication shows that machining marks were still visible in the leakage path, suggesting a low leakage rate during operation that led to minimal wastage of the RPV head.

10 REFERENCES

Bennetch JI, GE Modzelewski, LL Spain and GV Rao. 2002. "Root Cause Evaluation and Repair of Alloy 82/182 J-Groove Weld Cracking of Reactor Vessel Head Penetrations at North Anna Unit 2." In *2002 Proceedings of the ASME Pressure Vessels and Piping Conference (PVP2002), Service Experience and Failure Assessment Applications*, pp. 179-185. August 5-9, 2002, Vancouver, British Columbia, Canada. American Society of Mechanical Engineers, New York.

Clark AF. 1968. "Low Temperature Thermal Expansion of Some Metallic Alloys." *Cryogenics* 8(5):282-289.

Cumblidge SE, SR Doctor, GJ Schuster, RV Harris Jr., SL Crawford, RJ Seffens, MB Toloczko and SM Bruemmer. 2009. *Nondestructive and Destructive Examination Studies on Removed-from-Service Control Rod Drive Mechanism Penetrations*. NUREG/CR-6996, PNNL-18372, U.S. Nuclear Regulatory Commission, Washington, D.C.

Economou J, A Assice, F Cattant, J Salin and M Stindel. 1994. "NDE and Metallurgical Examination of Vessel Head Penetrations." In *3rd International Symposium on Contribution of Materials Investigation to the Resolution of Problems Encountered in Pressurized Water Reactors*. September 12-16, 1994, Fontevraud, France. French Nuclear Energy Society.

EPRI. 2005. *Materials Reliability Program: Destructive Examination of the North Anna 2 Reactor Pressure Vessel Head (MRP-142): Phase 1: Penetration Selection, Removal, Decontamination, Replication, and Nondestructive Examination*. EPRI Report 1007840, Electric Power Research Institute, Palo Alto, California.

EPRI. 2006. *Materials Reliability Program: Destructive Examination of the North Anna 2 Reactor Pressure Vessel Head (MRP-198): Phase 3: A Comparison of Nondestructive and Destructive Examination Findings for CRDM Penetration #54*. EPRI Report 1013414, Electric Power Research Institute, Palo Alto, California.

Gorman J, S Hunt, P Riccardella and GA White. 2009. "Chapter 44, PWR Reactor Vessel Alloy 600 Issues." In *Companion Guide to the ASME Boiler and Pressure Vessel Code, Volume 3, Third Edition*, ed: KR Rao. ASME Press, New York.

Grimmel B. 2005. *U.S. Plant Experience with Alloy 600 Cracking and Boric Acid Corrosion of Light-Water Reactor Pressure Vessel Materials*. NUREG-1823, U.S. Nuclear Regulatory Commission, Washington, D.C.

Hunt S and M Fleming. 2002. *Probability of Detecting Leaks in RPV Upper Head Nozzles by Visual Inspections, Revision 1, June 17, 2002*. Dominion Engineering, Inc., Reston, Virginia. Prepared for MRP PWR Alloy 600 Assessment Committee. U.S. Nuclear Regulatory Commission ADAMS Accession No. ML030860192.

IAEA. 2007. *Assessment and Management of Ageing of Major Nuclear Power Plant Components Important to Safety: PWR Pressure Vessel Internals, 2007 Update*. IAEA-TECDOC-1556, International Atomic Energy Agency (IAEA), Vienna, Austria.

Marquardt ED, JP Le and R Radebaugh. 2002. "Cryogenic Material Properties Database." In *Cryocoolers 11, 11th International Cryocooler Conference*, pp. 681-687. June 20-22, 2000, Keystone, Colorado. DOI 10.1007/0-306-47112-4_84. Springer US.

NRC. 2002. *Recent Experience with Degradation of Reactor Pressure Vessel Head*. Information Notice 2002-11, U.S. Nuclear Regulatory Commission, Washington, D.C. March 12, 2002. U.S. NRC Agencywide Data Access and Management System (ADAMS) Accession Number ML020700556.

APPENDIX A

PRECISION EDM NOTCH INFORMATION

APPENDIX A

PRECISION EDM NOTCH INFORMATION

Precision square-edged electrical discharge machining (EDM) notches were an essential aspect of the CRDM nozzle mockup specimen. As described in the calibration specimen design section, a variety of notches were chosen for the mockup specimen and allowed for a multitude of ultrasonic calibrations to be made. Understanding the phased-array probe resolution and detection characteristics allowed for a more thorough leak-path assessment to occur on North Anna Unit 2 removed-from-service Nozzle 63.

This appendix highlights the exact as-built dimensions and locations for all notches used in the mockup assembly specimen as provided by Western Professional, Inc., the EDM notch subcontractor. Page A-2 lists the as-built dimensions for the 16 EDM notches placed in the low-alloy steel material representing the RPV head. Page A-3 lists the as-built dimension for the 16 EDM notches place in the outer diameter of the Alloy 600 tube. Page A-4 shows the requested placement and size of the notches on the Alloy 600 tube outer diameter. Page A-5 shows the requested placement and size of notches 9–12. Page A-6 shows the requested placement and size of notches 13–16. Page A-7 shows the requested notch layout and sizing for the low-alloy steel material inner diameter.

WESTERN PROFESSIONAL, INC.
DBA WESTPRO LAB
3460 BRADY COURT NE
SALEM, OR 97303
(503)585-6263

Electrical Discharge Machining
Reference Standard Manufacturing
Optical Dimensioning System
Non Destructive Evaluation
RT, UT, MT, PT

CUSTOMER:	BATTELLE	STANDARD:	BLOCK STANDARD S/N 5381
DRAWING #:	CUSTOMER DRAWING	P.O. #:	135771
MATERIAL:	CARBON STEEL	SIZE:	4.102" Ø HOLE
DATE:	11-11-10	SPEC(S):	PER CUSTOMER DRAWING INSTRUCTIONS

BLOCK STANDARD S/N 5381

DEFECT DIMENSIONS (IN INCHES)					
NO.	DEPTH	LENGTH	WIDTH	LOCATION	ORIENTATION
1	.0011"	2.0017"	.0364"	I.D.	LONGITUDINAL
2	.0020"	2.0048"	.0379"	I.D.	LONGITUDINAL
3	.0029"	1.9974"	.0377"	I.D.	LONGITUDINAL
4	.0049"	2.0046"	.0372"	I.D.	LONGITUDINAL
5	.1000"	1.9891"	.0316"	I.D.	LONGITUDINAL
6	.1004"	1.9929"	.0624"	I.D.	LONGITUDINAL
7	.1007"	2.0067"	.1251"	I.D.	LONGITUDINAL
8	.1002"	1.9968"	.2514"	I.D.	LONGITUDINAL
9	.0786"	1.0061"	.0833"	I.D.	TRANSVERSE
10	.0789"	1.0011"	.0826"	I.D.	TRANSVERSE
11	.0768"	1.0015"	.0834"	I.D.	TRANSVERSE
12	.0811"	1.0004"	.0827"	I.D.	TRANSVERSE
13	.0811"	1.0009"	.0807"	I.D.	LONGITUDINAL
14	.0789"	1.0028"	.0821"	I.D.	LONGITUDINAL
15	.0805"	1.0011"	.0828"	I.D.	LONGITUDINAL
16	.0805"	1.0007"	.0810"	I.D.	LONGITUDINAL

SEE ATTACHED DRAWING FOR NOTCH LOCATIONS

NOTE: ALL DEPTH AND WIDTH MEASUREMENTS ARE BASED ON AN AVERAGE OF FOUR OR MORE READINGS.

ALL DIMENSIONS ARE MEASURED WITH DIMENSIONAL EQUIPMENT WHICH IS CERTIFIED AND TRACEABLE TO NIST #(708) #2343033 AND NIST (#783183) #3881145. NUCLEAR REGULATORY COMMISSION RULES AND REGULATIONS 10 CFR PART 21 APPLIES TO THIS ORDER. ALL NOTCHES MANUFACTURED PER WESTPRO PROCEDURE WQC-IV.

CERTIFIED BY: S. CHAMBERLAIN

APPROVED BY: _L.J. Chamberlain_

WESTERN PROFESSIONAL, INC.
DBA WESTPRO LAB
3460 BRADY COURT NE
SALEM, OR 97303
(503)585-6263

Electrical Discharge Machining
Reference Standard Manufacturing
Optical Dimensioning System
Non Destructive Evaluation
RT, UT, MT, PT

CUSTOMER:	BATTELLE	STANDARD:	PIPE STANDARD S/N 5382	
DRAWING #:	CUSTOMER DRAWING	P.O. #:	135771	
MATERIAL:	STAINLESS STEEL, HT# L215S	SIZE:	4.112" Ø X .6837" AWT	
DATE:	11-11-10	SPEC(S):	PER CUSTOMER DRAWING INSTRUCTIONS	

PIPE STANDARD S/N 5382

DEFECT DIMENSIONS (IN INCHES)				LOCATION	ORIENTATION
NO.	DEPTH	LENGTH	WIDTH		
1	.0011"	1.9976"	.0367"	O.D.	LONGITUDINAL
2	.0020"	1.9951"	.0366"	O.D.	LONGITUDINAL
3	.0030"	1.9971"	.0370"	O.D.	LONGITUDINAL
4	.0050"	1.9994"	.0371"	O.D.	LONGITUDINAL
5	.0980"	1.9945"	.0314"	O.D.	LONGITUDINAL
6	.1002"	2.0004"	.0632"	O.D.	LONGITUDINAL
7	.1012"	2.0066"	.1274"	O.D.	LONGITUDINAL
8	.1009"	1.9971"	.2526"	O.D.	LONGITUDINAL
9	.0795"	1.0007"	.0826"	O.D.	TRANSVERSE
10*	.0797"	1.0000"	.0851"	O.D.	TRANSVERSE
11	.0806"	1.0018"	.0807"	O.D.	TRANSVERSE
12	.0806"	1.0007"	.0822"	O.D.	TRANSVERSE
13	.0804"	1.0004"	.0818"	O.D.	LONGITUDINAL
14	.0840"	1.0015"	.0816"	O.D.	LONGITUDINAL
15	.0780"	1.0020"	.0818"	O.D.	LONGITUDINAL
16	.0809"	1.0012"	.0829"	O.D.	LONGITUDINAL

SEE ATTACHED DRAWING FOR NOTCH LOCATIONS

*NOTCH #10 WIDTH IS .001" OVER MAXIMUM TOLERANCE.

NOTE: ALL DEPTH AND WIDTH MEASUREMENTS ARE BASED ON AN AVERAGE OF FOUR OR MORE READINGS.

ALL DIMENSIONS ARE MEASURED WITH DIMENSIONAL EQUIPMENT WHICH IS CERTIFIED AND TRACEABLE TO NIST #(708) #2343033 AND NIST (#783183) #3881145. NUCLEAR REGULATORY COMMISSION RULES AND REGULATIONS 10 CFR PART 21 APPLIES TO THIS ORDER. ALL NOTCHES MANUFACTURED PER WESTPRO PROCEDURE WQC-IV.

CERTIFIED BY: S. CHAMBERLAIN

APPROVED BY: *L.J.Chamberlain*

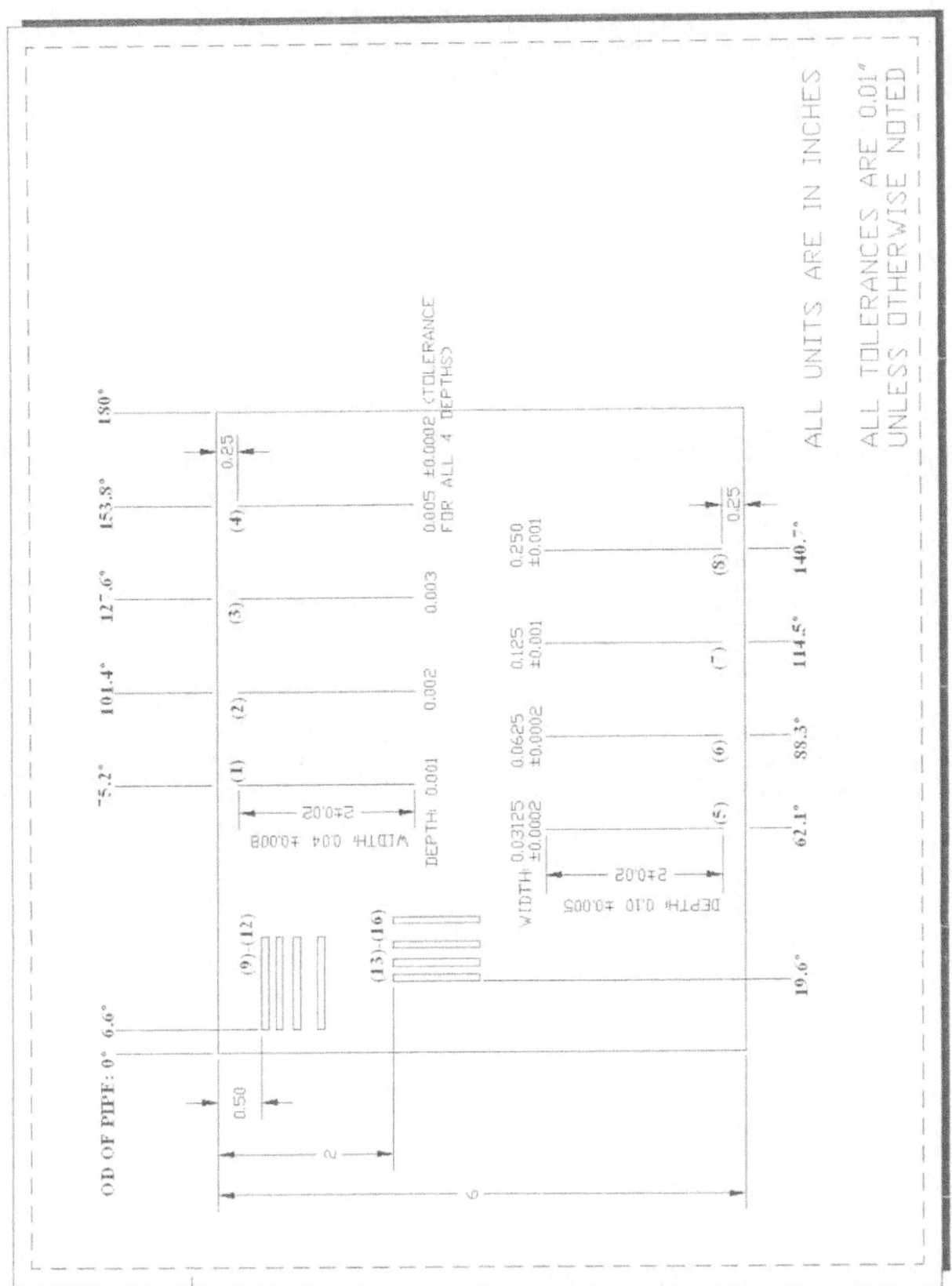

ALL UNITS ARE IN INCHES

ALL TOLERANCES ARE 0.01"
UNLESS OTHERWISE NOTED

A-4

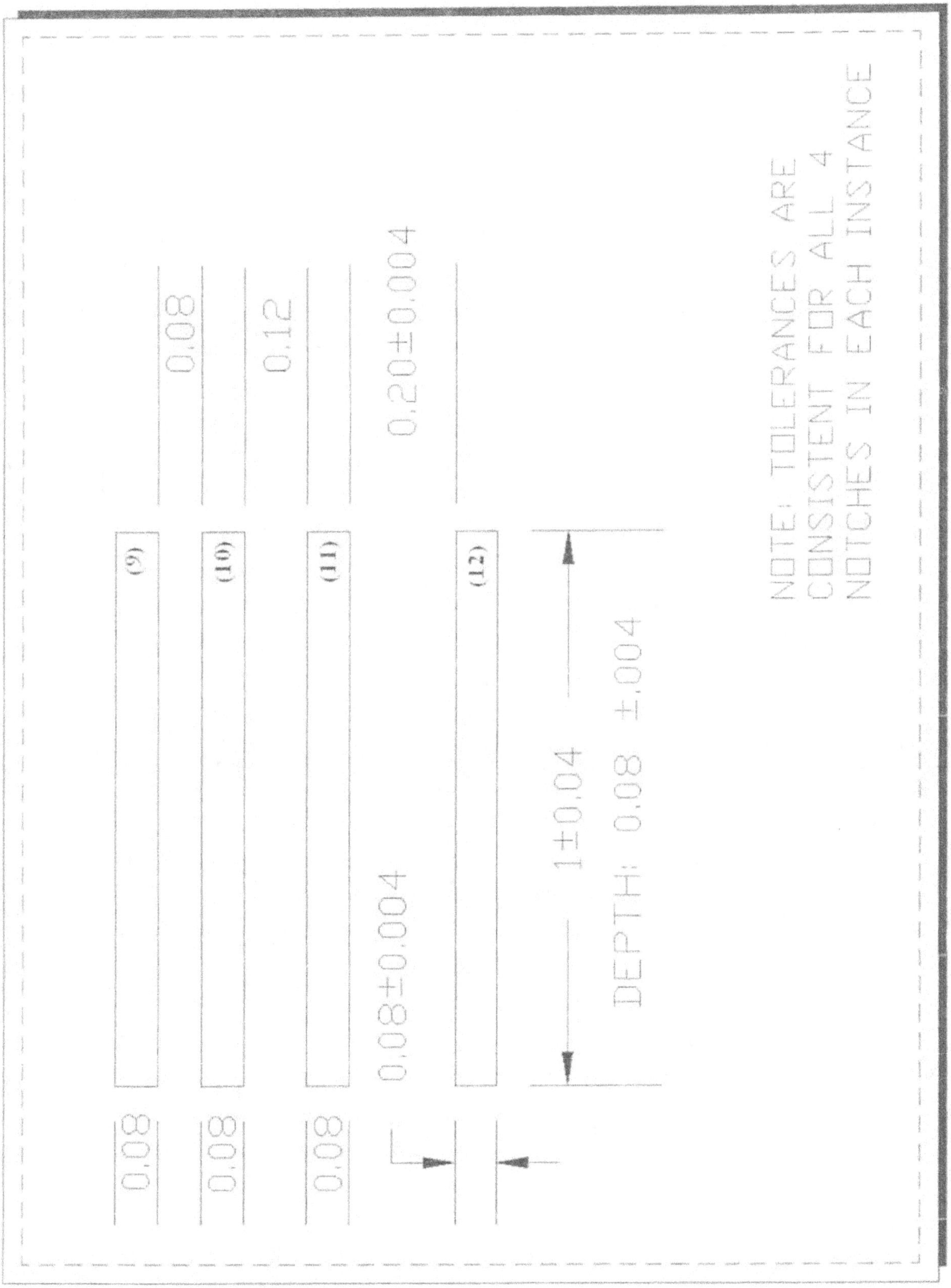

(9)

0.08

0.08

(10)

0.08

0.0

(11)

0.12

0.08±0.004

(12)

0.20±0.004

1±0.04

DEPTH: 0.08 ±.004

NOTE: TOLERANCES ARE
CONSISTENT FOR ALL 4
NOTCHES IN EACH INSTANCE

A-5

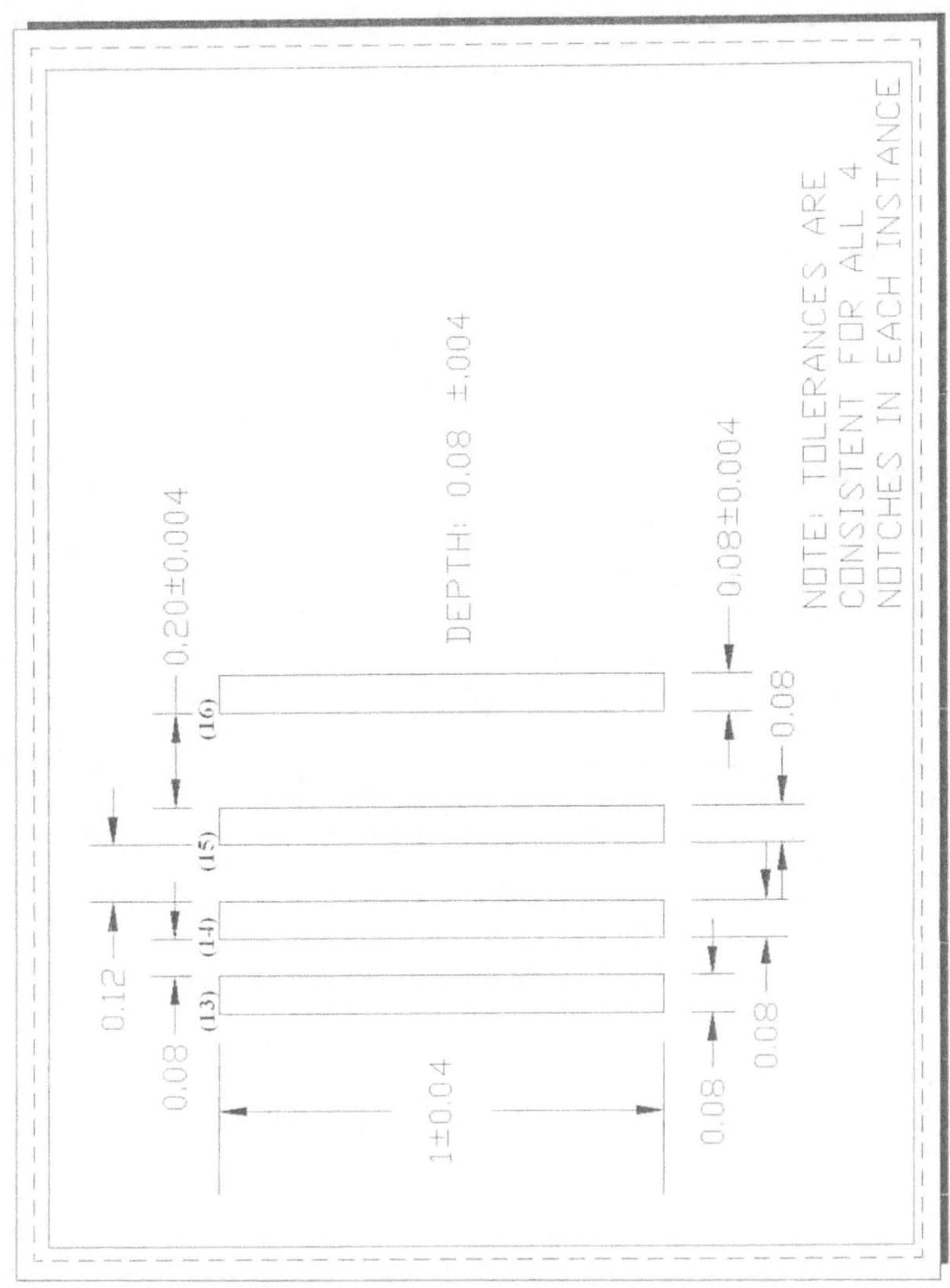

NOTE: TOLERANCES ARE CONSISTENT FOR ALL 4 NOTCHES IN EACH INSTANCE

DEPTH: 0.08 ±.004

0.20±0.004

0.08±0.004

0.08

(16)

(15)

(14)

(13)

0.12

0.08

1±0.04

0.08

0.08

0.08

0.0

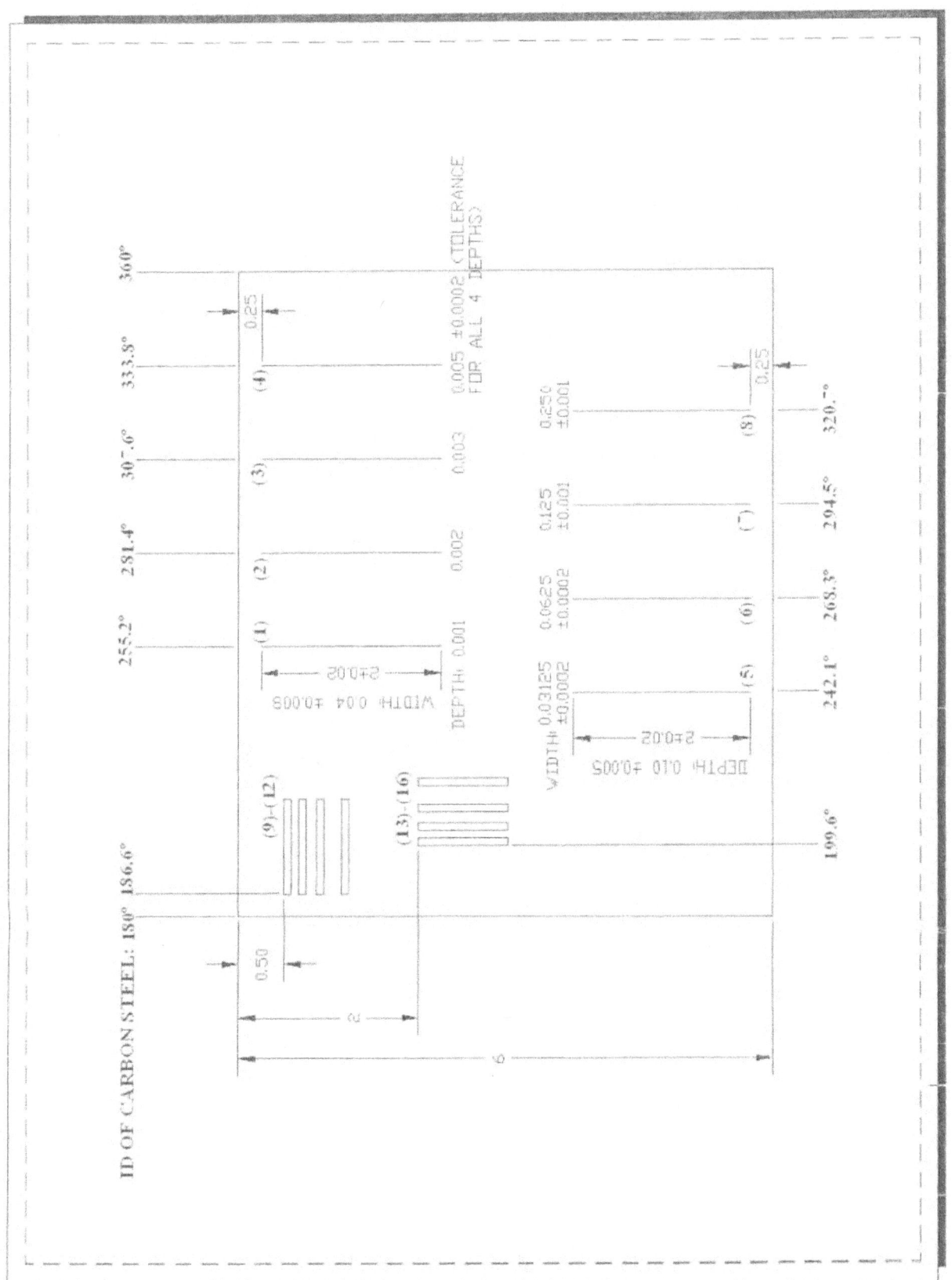

APPENDIX B

**THEORETICAL AND PRACTICAL INCONEL TUBE
SHRINKAGE**

APPENDIX B

THEORETICAL AND PRACTICAL INCONEL TUBE SHRINKAGE

To determine if the interference fit was attainable, the expected thermal shrinkage for the nozzle at liquid nitrogen temperature was calculated. Thermal expansion coefficients are generally given in tabular form for different materials over a particular temperature range. These table values assume a linear relationship over a limited temperature range. Cryogenic material properties were needed for the temperatures used in forming the interference fit. Two reference papers (Clark 1968; Marquardt et al. 2002) discussed material properties of metal alloys at cryogenic temperatures. Inconel 718 was one of the materials studied and represented the nozzle material for the purpose of material shrinkage calculations. To form the interference fit, the nozzle was taken from room temperature, 293°K (19.85°C, 67.73°F), to the temperature of liquid nitrogen, 77.2°K (−195.95°C, −320.71°F). Clark (1968) measured the thermal expansion coefficients from liquid hydrogen temperature to room temperature, 20°K (−253.15°C, −423.67°F) to 293°K (19.85°C, 67.73°F), in 10- or 20-degree steps for different metallic alloys. Results were presented in tabular form for the expansion or shrinkage relative to room temperature. From the table for Inconel 718 at 80°K:

$$223 = \left[\left(L_{293} - L_{80} \right) / L_{293} \right] \times 10^{5}$$

expansion relative to room temperature

(B.1)

where L_n = length at temperature n, in degrees K.

From Eq. (B.1), the shrinkage of the 104.14-mm (4.1-in.) diameter Alloy 600 tube at liquid nitrogen temperature relative to room temperature is calculated as approximately 0.232 mm (0.00914 in.).

Marquardt modeled the material properties over a large temperature range (4 to 300°K [−452.5 to 80.3°F or −269.2 to 26.9°C]) with a polynomial or a logarithmic polynomial equation. An equation for the integrated linear thermal expansion or shrinkage is given as:

$$\left(L_T - L_{293} \right) / L_{293} = \left(a + bT + cT^2 + dT^3 + eT^4 \right) \times 10^{-5}$$

(B.2)

The coefficients for 718 Inconel are listed as:

a = −2.366E+02
b = −2.218E-01
c = 5.601E-03
d = −7.164E-06
e = 0

From Eq. (B.2), the shrinkage of the tube is calculated as approximately 0.233 mm (0.00917 in.) at 77K. This result very closely matches the result from Eq. (B.1) given above.

Prior to assembly of the mockup, experimental verification of the calculated shrinkage was conducted on a section of Alloy 600 tubing cut from a tube similar to that which was used for the mockup. At room temperature, five outside diameter measurements were acquired at 0–180 and five at 90–270 degrees using a calibrated caliper, shown in Figure B.1, at marked locations along the axial length of the specimen. Temperature measurements were obtained using a Raytek ST80XXUS infrared (IR) standoff thermometer. At room temperature the average diameter of the tube was 104.902 mm (4.130 in.). The tube section was then submerged in a liquid nitrogen (LN) bath for approximately 2.5 minutes. The chilled tube section was promptly removed from the LN bath and the diameter of the tube was again measured at the same locations. The temperature of the tube upon removal from the LN bath was unattainable as it surpassed the low end capability of the IR thermometer (−32 to 760 degrees C [−25 to 1400 degrees F]). The theoretical temperature of LN is −195.95 degrees C (−320.71 degrees F). At cold temperatures, the average diameter of the tube was 104.699 mm (4.122 in.). The tube section was allowed to re-equilibrate to room temperature and additional diameter measurements were made. These showed that the average nozzle diameter returned to 104.902 mm (4.130 in.). A full set of diameter data can be viewed in Table B.1, where data were acquired in descending order from position 5 to 1. Figure B.2 shows the tube diameter at the three stages of this experiment as a function of axial position: room temperature initial (RT initial), after LN, and room temperature final (RT final). The results from this test indicated that the Alloy 600 tube material shrank an average of 0.203 mm (8 mils) in diameter when chilled with LN and then was restored to its original size after returning to room temperature. The CRDM calibration specimens were designed and machined for a 0.076-mm (3-mils) diameter interference fit. Tube shrinkage of an additional 0.127 mm (5 mils) provided the necessary room for slipping the machined low-alloy steel blocks over the tube during mockup assembly. Moreover, it was equally important that the tube return to its original size (at room temperature) in order to create the tight interference fit. This successful preliminary test proved that both necessary requirements could be achieved.

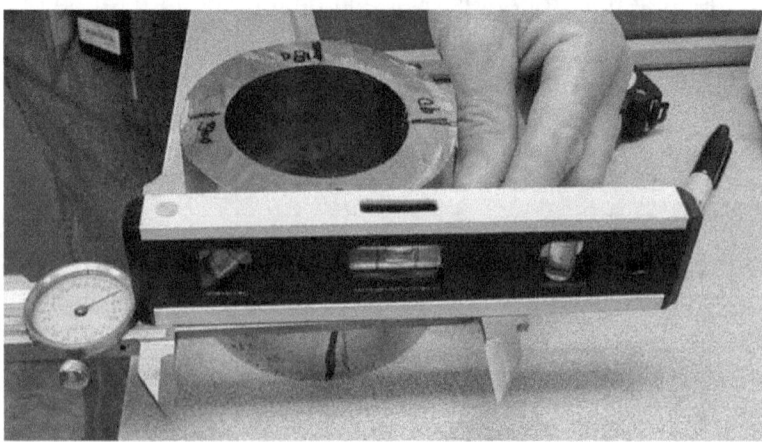

Figure B.1 Diameter Measurements Acquired at Room Temperature Using a Caliper

Table B.1 Alloy 600 Tube Diameter Measurements Verses Temperature

		Room Temp 73.9		Liquid Nitrogen Unknown			Return to Room Temp 73.9	
Temp (F)		0–180	90–270	0–180	90–270		0–180	90–270
Diameter (in.)	1	4.130	4.130	4.122	4.125		4.130	4.130
	2	4.130	4.130	4.122	4.122		4.130	4.130
	3	4.130	4.130	4.124	4.122		4.130	4.130
	4	4.130	4.130	4.120	4.122	Λ	4.130	4.129
	5	4.130	4.130	4.120	4.122	t=0	4.130	4.129
Average:		4.130		4.122			4.130	

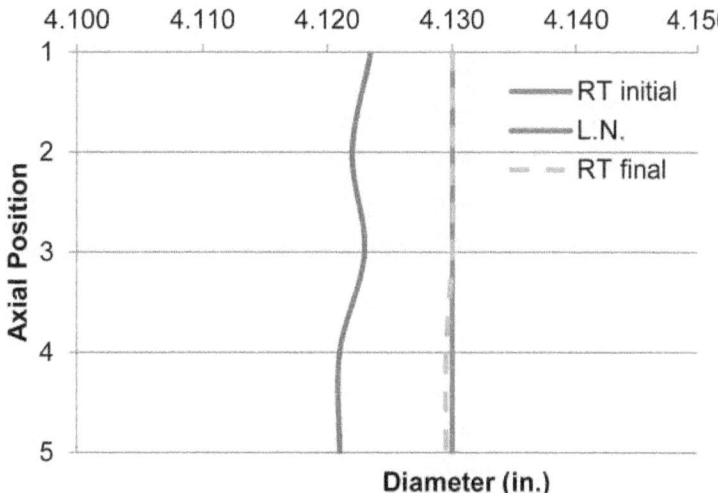

Figure B.2 Tube Shrinkage Measurements

APPENDIX C

SUPPLEMENTAL INFORMATION ON DESTRUCTIVE CUTTING

APPENDIX C

SUPPLEMENTAL INFORMATION ON DESTRUCTIVE CUTTING

B&W used an industrial 5.79-m (19-ft) CobraFab Industries, Inc. band saw model VH2532HD with 'Cobra Strike' and an interchangeable blade option for all cutting purposes associated with the destructive activity. This saw was equipped with a ±45° hydraulic miter attachment and was capable of cutting pieces measuring 63.5 cm (25 in.) wide by 81.3 cm (32 in.) high in the vertical position and 63.5 cm (25 in.) wide by 55.9 cm (22 in.) high in the ±45° positions. A total of six 5.79 m (19-ft) blades were acquired from Scarney Industries, Youngstown, Ohio. Two coarse tooth (2-3 teeth per inch) and three fine tooth (4-6 teeth per inch) bimetallic blades with positive rake, variable pitch, and variable tooth were used. Additionally, one carbide grit-abrasive blade was also used. In general, coarse tooth blades were used for thicker section cutting, 20.3 cm (8 in.) and larger, and the fine toothed blades were used for precision cutting material under 20.3 cm (8 in.) in thickness. The carbide grit abrasive blade was used when the bimetallic blades were ineffective. Figure C.1 shows the three blade types used. All cuts in this activity were conducted without lubrication or coolant to minimize disturbance to the interference fit region.

Figure C.1   Three Blade Types were Utilized to Section the RPV Head and CRDM Nozzle. From left: fine tooth blade, coarse tooth blade, and carbide grit blade.

The initial size reduction cuts were performed on the Nozzle 63 specimen using one of the coarse toothed blades prior to the arrival of PNNL staff. Non-essential material was removed to reduce weight and to facilitate proper blade placement on the specimen during critical cuts. Figure C.2 shows images from the size reduction activity. All cuts in this activity were conducted without lubrication or coolant to minimize disturbance to the interference fit region.

Figure C.2 Size Reduction Cutting Activity

After size reduction, the nozzle assembly was prepared for the dissection cut that separated the high and low sides of the assembly. The cut line was selected to start at approximately the 95-degree mark (Figure C.3), and follow through to the 275-degree mark. The line placement was based on the ultrasonic data and chosen to preserve the primary leak path previously identified in the ultrasonic images.

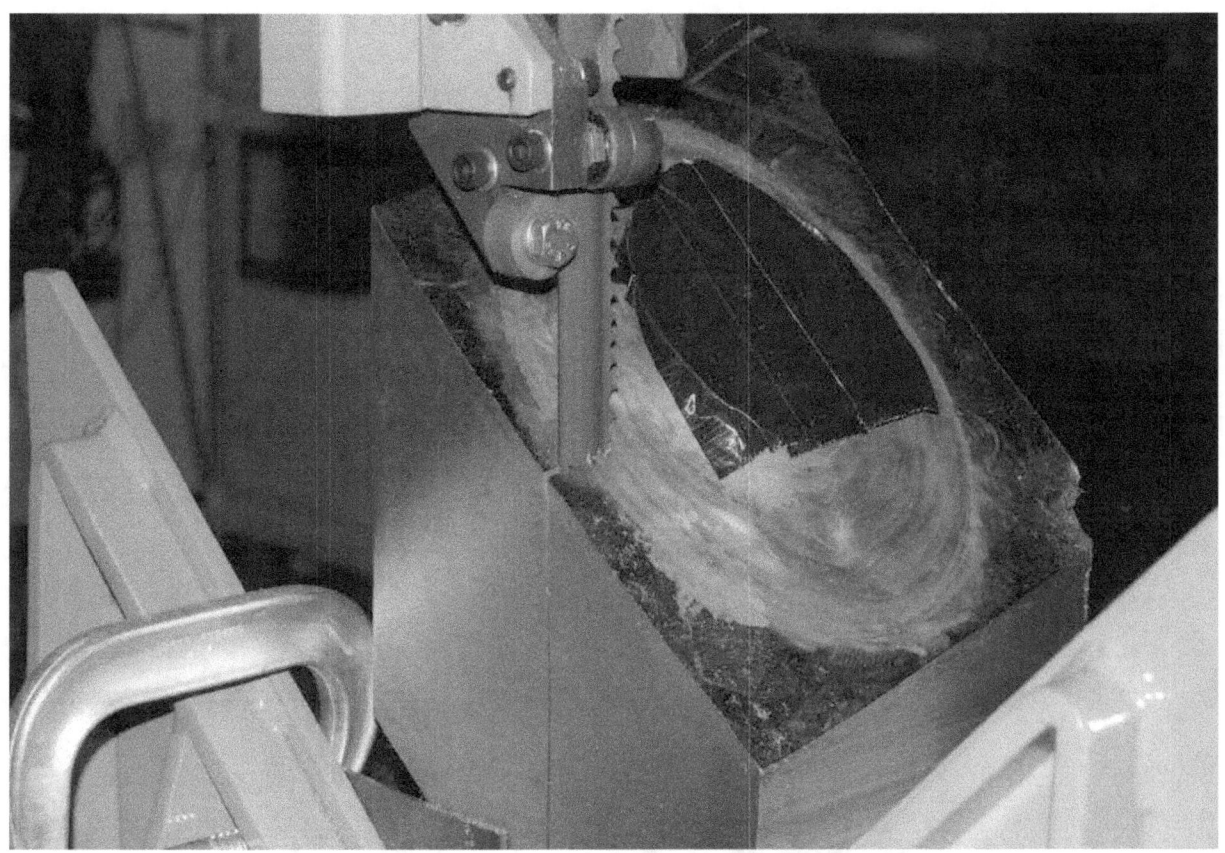

Figure C.3 Start of the Dissection Cut

Dissection cutting began with the coarse-toothed blade. The cut progressed for approximately 20 minutes until an unidentified 'hard spot' was reached at the outer edge of the Alloy 600 nozzle region and stopped the cutting progression. A slower feed rate with a faster blade speed was attempted, but did not traverse the hard spot. The cause of the hard spot was not fully investigated, but a likely cause was from a cold-worked region of material within the heat-affected zone of the J grove weld adjacent to the nozzle outer diameter (OD). The coarse-cut blade was exchanged with a fine-toothed cutting blade and the cut was attempted again without success. Finally, the carbide blade was employed to abrasively grind through the hard spot. This blade required a greatly reduced feed rate, thus lengthened the cutting time. Further, the kerf of the carbide blade was thicker than the cutting blades, requiring the cut to be restarted at the initial cut path. The carbide grit abrasive blade is shown in Figure C.4. The cut through this hard spot required approximately 2 hours to complete.

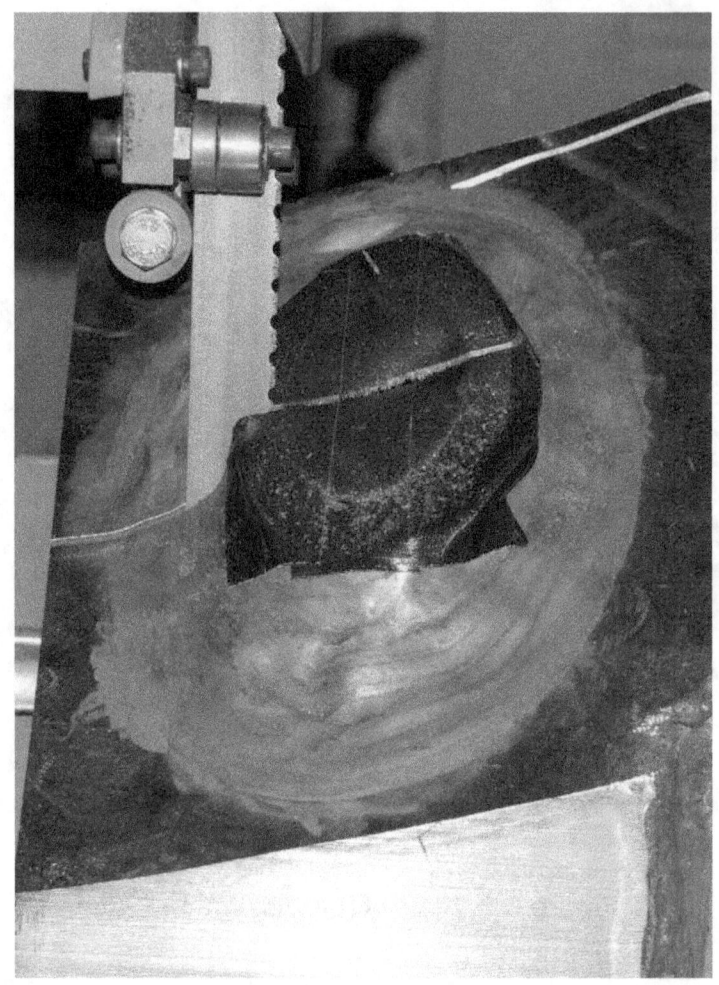

Figure C.4 Abrasive Carbide Blade Progressing Through the Hard Spot

After cutting through the first side of the nozzle, the carbide blade was exchanged with the fine-toothed cutting blade and the dissection cutting continued. Another hard spot was incurred in the J-groove weld region. Again, the carbide blade was used to cut through the hard spot with a slow feed rate. Finally, the fine-toothed blade was re-engaged and the dissection cut completed. From the exposed surfaces, the triple point was identified along with the weld and butter regions as shown in Figure C.5.

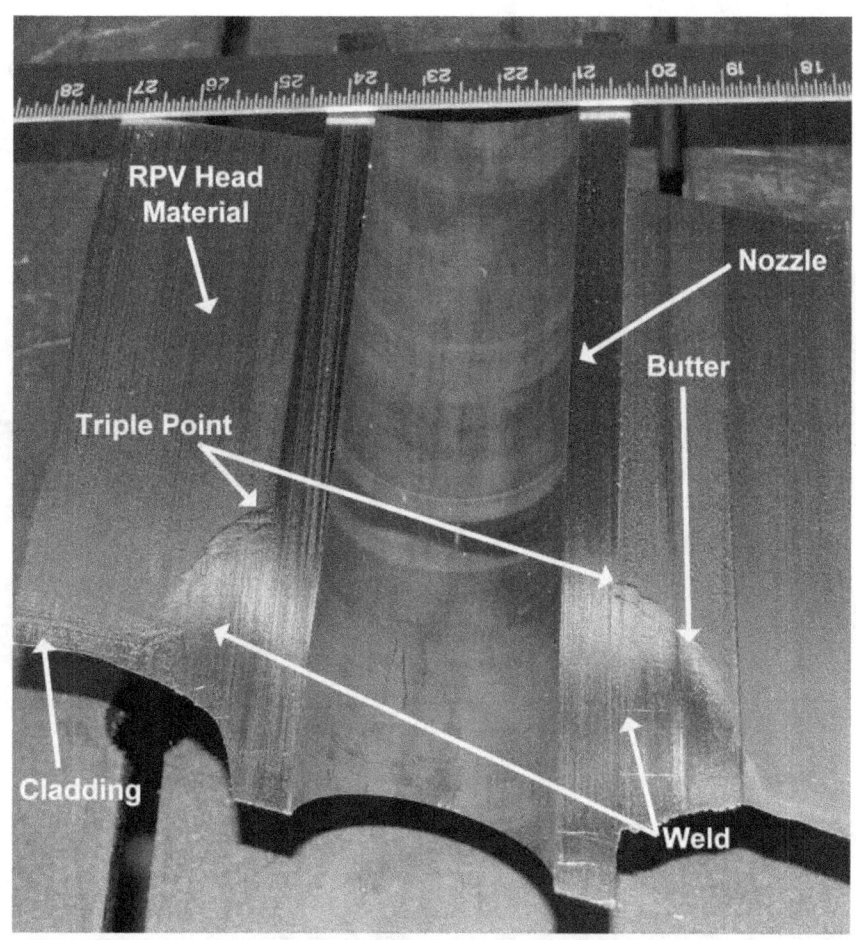

Figure C.5 Nozzle 63 Assembly Cut in Half by Dissection Cut

The cut to remove the J-groove weld was made 6.35 mm (0.25 in.) above the butter/triple point region in the reactor pressure vessel (RPV) head as seen in Figure C.6. The high or uphill side half-portion was selected for the first cutting. As previously stated, the 'high' side has several potential leak paths whereas the 'low' or downhill side had the primary leak path as identified in the ultrasonic data. The specimen was secured and the band saw tilted to an approximate 43-degree angle to match the angle between the nozzle and head as shown in Figure C.7. Cutting at this angle maximized the annulus region that was exposed while keeping the weld and butter regions intact for future evaluation. As the cut was designed to pass only through the low-alloy RPV material, the fine toothed cutting blade was selected for use. During the J-groove weld removal cut, another hard spot was encountered near the outside of the nozzle. Attempts were made to continue cutting with the fine-toothed blade until abrasive wear on one side of the blade resulted in the cut veering away from the desired cut line. Blades were exchanged and the carbide blade was used to cut through the hard spot and also to finish the cut. Before the cut broke through the inside of the tube, a vacuum equipped with a high-efficiency particulate air (HEPA) filter was added to collect and capture any radioactive oxide particles that were discharged from the cut as pictured in Figure C.8.

Figure C.6 J-groove Weld Removal Cut Line Placement

Figure C.7 J-groove Weld Cut on the High Side

Figure C.8 Abrasive Blade Cutting Through the Nozzle with a HEPA Vacuum Nozzle

The cutting continued until the entire nozzle was separated from the J-groove weld region as pictured in Figure C.9. At this point, the nozzle region above the weld was freely released from the RPV head material. Removal of the nozzle exposed the annulus region of the high-side section as shown in Figure C.10. At this point all available blades were exhausted. Replacement blades were ordered to finish the cut and complete removal of the J-groove weld region. A Nikon D40x camera was used to acquire high-resolution photographic documentation of the annulus region. This activity was provided by B&W. A subsequent cut was conducted on the low-side section to expose its annulus region containing the primary leak path (Figure C.11). The nozzle freely released from this portion as well.

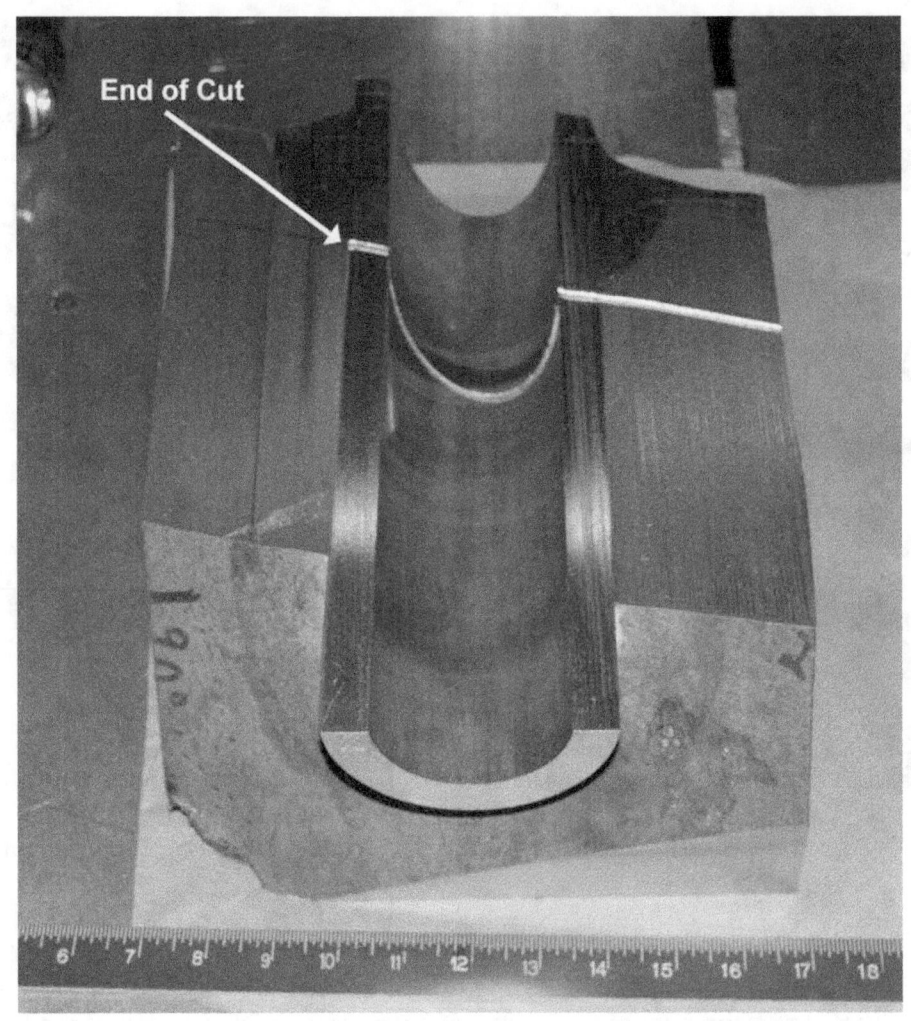

Figure C.9 End of J-groove Weld Removal Cut

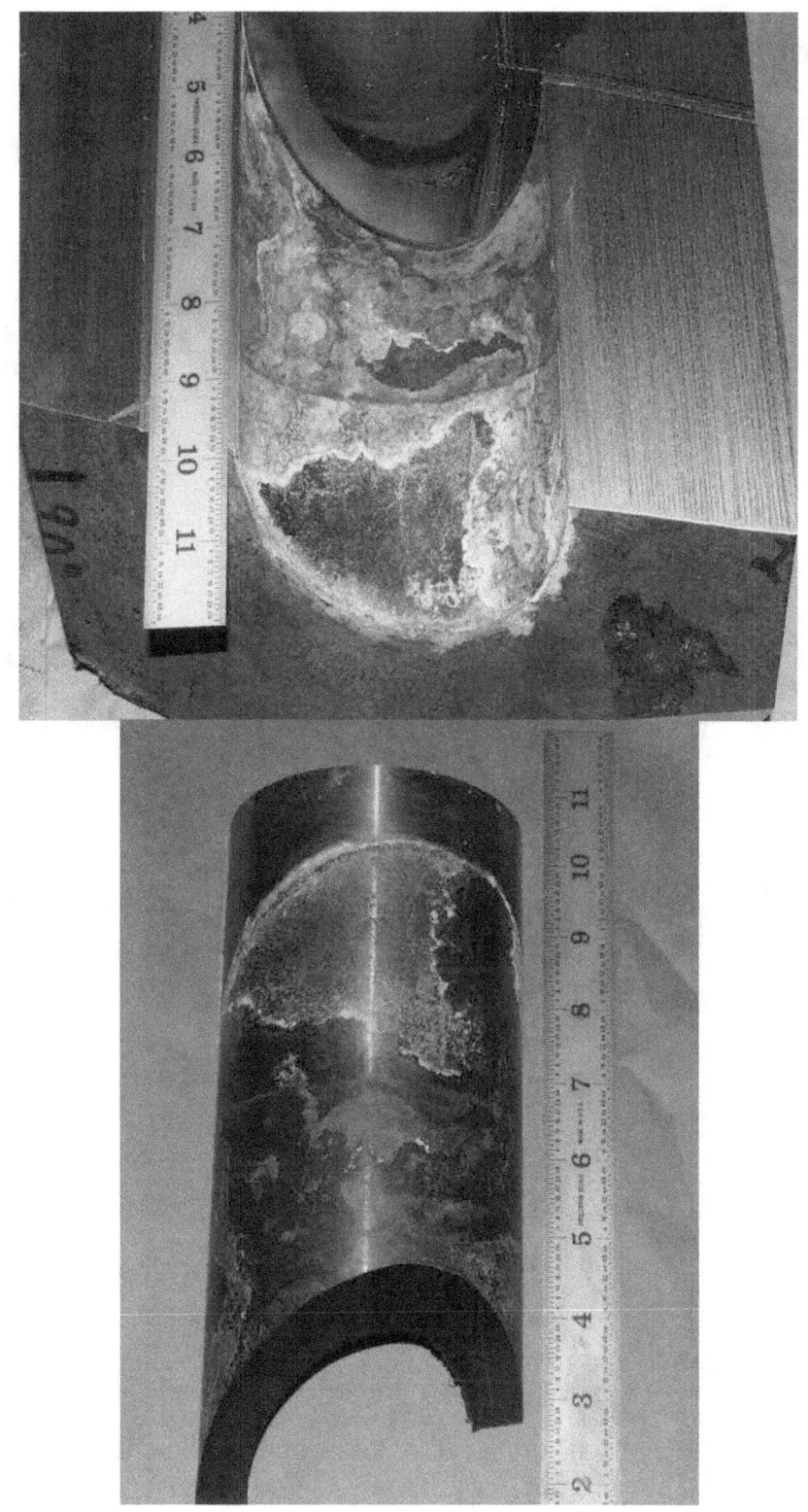

Figure C.10 Exposed RPV Head and Nozzle from High Side Section

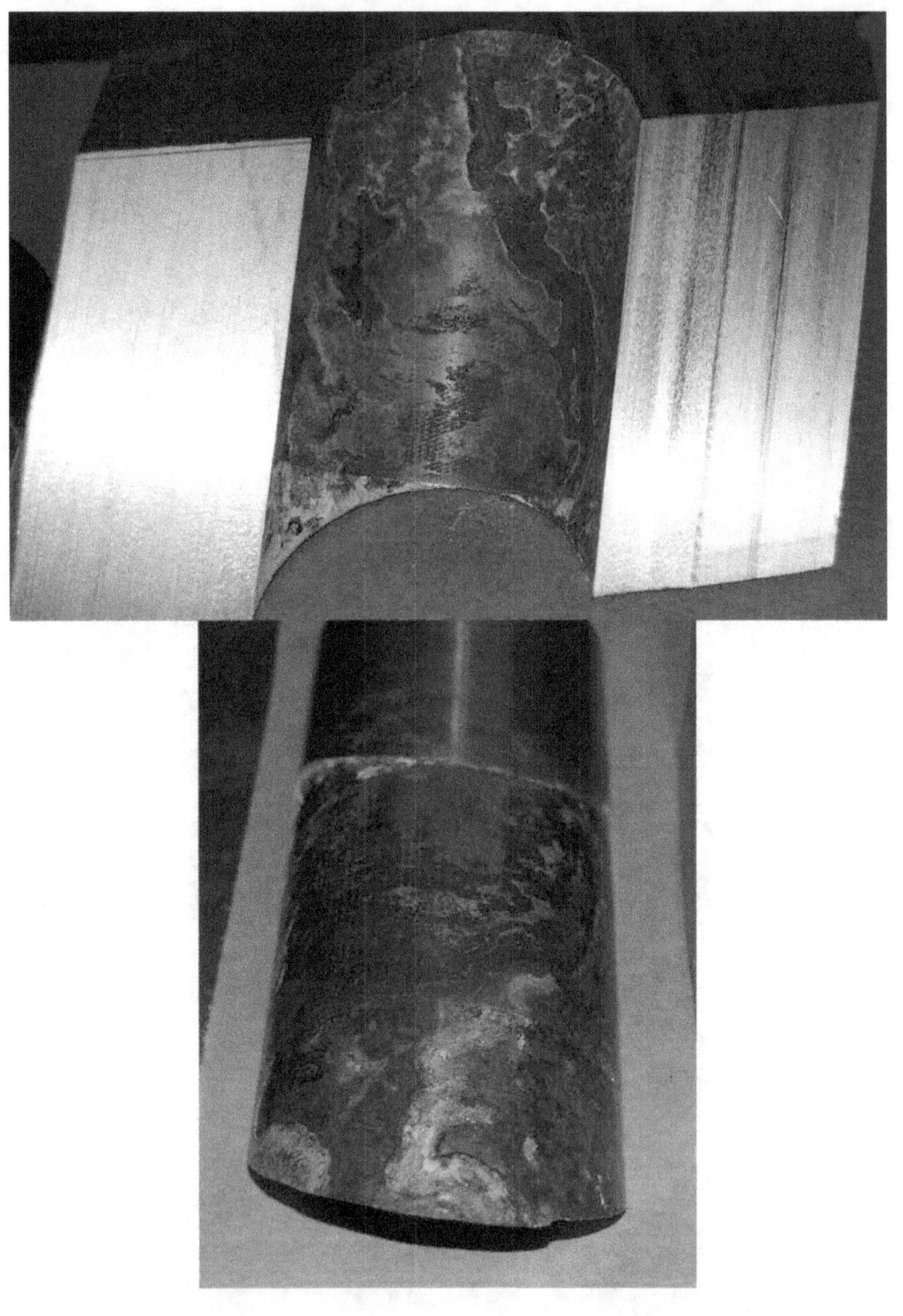

Figure C.11 Exposed RPV Head and Nozzle from Low-Side Section

NRC FORM 335 (12-2010) NRCMD 3.7	U.S. NUCLEAR REGULATORY COMMISSION	1. REPORT NUMBER (Assigned by NRC, Add Vol., Supp., Rev., and Addendum Numbers, If any.)
	BIBLIOGRAPHIC DATA SHEET *(See instructions on the reverse)*	NUREG/CR-7142

2. TITLE AND SUBTITLE	3. DATE REPORT PUBLISHED	
Ultrasonic Phased Array Assessment of the Interference Fit and Leak Path of the North Anna Unit 2 Control Rod Drive Mechanism Nozzle 63 with Destructive Validation	MONTH August	YEAR 2012
	4. FIN OR GRANT NUMBER N6783	
5. AUTHOR(S) S. L. Crawford, A. D. Cinson, P. J. MacFarlan, B. D. Hanson, R. A. Mathews	6. TYPE OF REPORT Technical	
	7. PERIOD COVERED (Inclusive Dates) 05/03/2010-07/31/2012	

8. PERFORMING ORGANIZATION - NAME AND ADDRESS (If NRC, provide Division, Office or Region, U. S. Nuclear Regulatory Commission, and mailing address; if contractor, provide name and mailing address.)

Pacific Northwest National Laboratory
P. O. Box 999
Richland, WA 99352

9. SPONSORING ORGANIZATION - NAME AND ADDRESS (If NRC, type "Same as above", if contractor, provide NRC Division, Office or Region, U. S. Nuclear Regulatory Commission, and mailing address.)

Division of Engineering
Office of Nuclear Regulatory Research
U.S. Nuclear Regulatory Commission
Washington, D.C. 20555-0001

10. SUPPLEMENTARY NOTES

11. ABSTRACT (200 words or less)

The objective of this investigation was to evaluate the efficacy of ultrasonic testing (UT) for primary water leak path assessments of reactor pressure vessel (RPV) upper head penetrations. Operating reactors have experienced leakage when stress corrosion cracking of nickel-based alloy penetrations allowed primary water into the annulus of the interference fit between the penetration and the low-alloy steel RPV head. In this investigation, UT leak path data were acquired for an Alloy 600 control rod drive mechanism nozzle penetration, referred to as Nozzle 63, which was removed from the North Anna Unit 2 reactor when the RPV head was replaced in 2002. In-service inspection prior to the head replacement indicated that Nozzle 63 had a probable leakage path through the interference fit region. Nozzle 63 was examined using a phased-array UT probe with a 5.0-MHz, eight-element annular array. Immersion data were acquired from the nozzle inner diameter surface. The UT data were interpreted by comparing to responses measured on a mockup penetration with known features. Following acquisition of the UT data, Nozzle 63 was destructively examined to determine if the features identified in the UT examination, including leakage paths and crystalline boric acid deposits, could be visually confirmed. Additional measurements of boric acid deposit thickness and low-alloy steel wastage were made to assess how these factors affect the UT response. The implications of these findings for interpreting UT leak path data are described.

12. KEY WORDS/DESCRIPTORS (List words or phrases that will assist researchers in locating the report.)	13. AVAILABILITY STATEMENT
Ultrasonic phased array evaluation, ultrasonic testing, UT, control rod drive mechanism, CRDM, reactor pressure vessel, RPV, primary water stress corrosion cracking, PWSCC, interference fit, J-groove weld, leak path, Alloy 600, boric acid deposits, nondestructive evaluation, NDE, North Anna Unit 2, Nozzle 63, pressurized water reactor, PWR	unlimited
	14. SECURITY CLASSIFICATION
	(This Page) unclassified
	(This Report) unclassified
	15. NUMBER OF PAGES
	16. PRICE

NUREG/CR-7142

Ultrasonic Phased Array Assessment of the Interference Fit and Leak Path of the North Anna Unit 2 Control Rod Drive Mechanism Nozzle 63 with Destructive Validation

August 2012

www.ingramcontent.com/pod-product-compliance
Lightning Source LLC
Chambersburg PA
CBHW080257180526

45167CB00006B/2565